Berry's Golden Rule Poultry Farm
or; Profitable Poultry

by Mrs. A.A. Berry

with an introduction by Jackson Chambers

This work contains material that was originally published in 1913.

This publication is within the Public Domain.

This edition is reprinted for educational purposes
and in accordance with all applicable Federal Laws.

Introduction Copyright 2018 by Jackson Chambers

The World's Largest Selection of Vintage Poultry Books

www.VintagePoultry.com

Self Reliance Books

Get more historic titles on animal and stock breeding, gardening and old fashioned skills by visiting us at:

http://selfreliancebooks.blogspot.com/

Introduction

I am pleased to present yet another title on Poultry.

The work is in the Public Domain and is re-printed here in accordance with Federal Laws.

As with all reprinted books of this age that are intended to perfectly reproduce the original edition, considerable pains and effort had to be undertaken to correct fading and sometimes outright damage to existing proofs of this title. At times, this task is quite monumental, requiring an almost total "rebuilding" of some pages from digital proofs of multiple copies. Despite this, imperfections still sometimes exist in the final proof and may detract from the visual appearance of the text.

I hope you enjoy reading this book as much as I enjoyed making it available to readers again.

Jackson Chambers

From a Photograph Showing a Bird's Eye View of the Most Wonderful Poultry Farm in the World

MY DEAR FRIEND:—Now I am going to write a pretty long introduction; but I want to know all the people who have sent for this book and I want you all to know me WELL. Know me well enough to have confidence in me that I will do just what I say I will and give you a good square deal. So I want you to read this through.

The old saying, "Nothing succeeds like success," is especially true in the poultry business, and I do not know of any business that so many succeed so easily or well with as little capital as those engaged in this line of work.

This is the seventh edition of my book, Profitable Poultry, and each one is a little better than the former. And it should be, as I add a lot of experience each year. I hope you will like this copy.

I have been agreeably surprised at the success we have attained in this business at the Golden Rule Farm. The photo on the opposite page will give you some idea of the magnitude of our establishment. This is from a real photo—not some artist's imagination, as many of the pictures showing poultry farms are.

The farm is my home; it contains over 100 acres and is devoted to poultry and growing of seeds.

I have been interested in poultry ever since I was a little girl. I was born in a log cabin on the frontier in the pioneering days in southwestern Iowa. I can remember my parents taking eggs to town and selling them for four cents per doz. So you see it was not so profitable as now.

When I was a little girl my mother gave me a few setting hens and eggs enough to put under them. I succeeded so well that I was given great encouragement and words of praise; thus my start in the poultry business began. I have simply grown up in it. I taught school a few years, so have had a very busy life.

I have four dear little children going to school. The junior members all take a great interest in the poultry business and our work and assist in many ways. These little olive plants are the things we work for and make life worth living; make a happy and interesting home.

Who would know more about the poultry business than a successful woman, and are women not more successful in the poultry business than men? I do not think that any one will deny the statement; that nine-tenths of the poultry raised in the United States is raised by women, and this assertion is made in all due reverence to man as the provider of the home. The chickens have been left to the women folks to look after. In some things women excel, and that is in raising and caring for babies and chicks. They know what is best for them, they know how to develop them for best results.

I know how—I know good poultry—know how to construct the best kind of an incubator and brooder and know the best of all kinds of poultry supplies. I have found that out by study and EXPERIENCE, which is the best teacher.

I have the best and can furnish you with the best of all these things. If I did not know these things—if I did not keep the best, I could not have made the great success that I have made.

Now I can help you to succeed as I have done and have helped others to succeed. Please give me a chance. Write to me fully as to your wants, location, condition, etc. I will take pleasure in assisting you in every way possible. And you cannot make a better start than to send me an order for what you want.

Read what I have to say in this book. You will find a great deal of information that will be of value to you. I have tried to make it practical, and a great many who have read it were kind enough to write and praise it very much.

A woman is desperately in earnest and will make her word good. I certainly will mine and give you a square deal, with you at the best end of it. Give me a chance to serve you and you will never regret it.

To My Old Customers

I send a special greeting. I know that you have been pleased in the manner in which I have handled your order. I have received thousands of nice letters from you for which you have my heartiest thanks.

I am more than encouraged by the kind words of praise and cherished letters you have sent me.

The Golden Rule is our motto and all deals are made by "doing unto others as we would have others do unto us;" so is there any reason for us not making a success and our customers and friends making the poultry business pay and a success?

To all my friends in the poultry business, to present and prospective customers, I dedicate this book, "Profitable Poultry."

Best wishes, long life, prosperity and successful poultry business to all, I remain,   Your true friend,
                                                                         MRS. A. A. BERRY.

P. S.—Be friendly; write me.

THE JUNIOR MEMBERS.

These are the junior members of the firm. Ethel and Anna are feeding their favorite flocks of White Leghorns. Ethel, who is now 18 years old, stands to the right; Anna, now 17 years old, just looked up as I took the picture.

Below, the two boys are shown as they were leaving the White Leghorn pen with a basket of eggs; looks rather risky to let such small boys handle a half bushel basket of valuable eggs, but they are trained to be careful.

Ernest, to the left, is now 14 years old, while George is now 10 years old, but such a husky lad that he is almost as large as Ernest. I took this picture and just before I exposed the plate I told Ernest that he would have to get a new pair of overalls, as his bare legs would show, or get a mamma who had time to sew them up; that is what makes the boys laugh.

PROFITABLE POULTRY

By Mrs. A. A. Berry.

WILL IT PAY? This question first arises to one contemplating the poultry business: Does it pay? Can I make a living or make money in the poultry business? We would emphatically say: Yes, you can both make a living and lay up money besides. We are backed in our answer by the hundreds of profitable poultry plants devoted to the raising of broilers for early market, by mammoth farms devoted to the production of eggs, by the large plants on which "soft roasters" are produced, by the mammoth establishments devoted to pure bred fowls and thousands of smaller poultry farms which produce both eggs and chickens for the market, and by the magnificent pure bred breeding plants that are scattered all over the country.

Thousands of farmers (often their wives) are raising enough chickens, the profit on which pays the grocer's bill as well as the dry goods and clothing bill and meets all incidental expenses.

There is nothing on the farm that will make the good clear money that poultry will, and there are thousands of poultry raisers who will testify to the truth of this assertion.

RAISING CHICKENS IS AMERICA'S LARGEST INDUSTRY.

There is no greater money maker in the world today than the patient American hen. Nearly forty-five million crates of eggs are sold yearly. According to the government's reports, "eggs and poultry earnings for one year over six hundred and forty-five millions of dollars." Such gigantic figures are not readily understood only by comparing them with other industries. Study this illustration:

HOW UNCLE SAM'S LARGEST INDUSTRY COMPARES WITH SOME OTHERS.

This diagram shows the poultry and egg output of the United States in comparison with seven other industries. Their value is given as follows:

```
Poultry and eggs .......................... $645,421,216
Gold, Silver, Sheep and Wool..............    391,564,432
Cotton ....................................   362,863,456
Wheat .....................................   329,734,862
Hogs ......................................   289,673,213
Oats ......................................   120,466,004
Potatoes ..................................    91,300,000
Tobacco ...................................    60,290,000
```

Can you grasp it? Does it not make it plain that the poultry industry is away ahead of all others and greater than a number of the supposedly large ones? The fact that the poultry business is of such mammoth proportions proves beyond the shadow of a doubt that it is highly profitable, as it is not characteristic of the American people to carry such a stupendous business at a financial loss from year to year. Then again, to any one familiar with the advertising columns of the poultry journals of our country, to say nothing of the thousands of advertisements of poultry that appear in stock, agricultural, fruit and other journals, it is evident there must be some money in the business of raising pure bred poultry.

In view of the above facts, we are warranted in the statements that the farmer who gives his poultry the same care and attention that he does the other domestic animals and other departments of his farm, will soon wake up to the fact that there is nothing on his farm that gives as good returns from a financial point of view for the capital invested and time spent, as the poultry. To be sure, it is considered by many as a small business and only fit for the women folks, and beneath the farmer's business dignity to fool with such small things as chickens and eggs. But the above figures do not show that it is a small thing and we do not hesitate to say to one who is looking for a paying business from a small investment of capital that we know he will find it in profitable poultry.

Will it last?

Is there not danger of overdoing the chicken business? No, it is the most likely business in America today. The consumption of eggs and poultry is something enormous. The United States cannot supply the great demand and we are compelled to import from other nations at the rate of thousands of cases of eggs per year. We get large quantities from our neighbors in Canada and a great many come from across the sea. We simply cannot supply the demand even at the high price that has prevailed for the past few years, in many states the farmer obtaining 20c to 40c per dozen during the spring and summer and double that price in the winter time.

There is a great demand for better bred poultry, as good pure bred stock of all kinds will pay better and be more satisfactory in every way.

There are a great many advantages in a pure bred flock of chickens, ducks or geese, over that of dunghills. Pure breds lay more eggs, are more uniform and of better quality. They will make more weight, thus being worth more when sold. By all means raise the best. Don't be content with "dunghills." Get some pure breds either by procuring some eggs, a few birds of both sexes, or some good males to head your flocks.

There is no danger of overdoing the poultry business.

THERE IS MONEY IN EGGS AND POULTRY.

When contemplating going in to a business of any description, the question naturally arises: Is there any money in it? And a thorough investigation is made with that one object in view.

Now, when we say that there is money in eggs and poultry, it is not an utterance from a phonograph, but backed by human intellect and experience. I do not make the assertion because some one else has told me, and thus believed it and thought there ought to be money in it; or figured it out on paper; but I positively know it to be a fact.

What other one thing is there such a demand for, as poultry and eggs? In fact, the demand is so great that prices for eggs always range from 15 cents to 30 cents per dozen and in many localities as high as 60 cents. This in itself shows that a good flock of hens is as good as a U. S. mint.

How many a family is supplied with daily food and wearing apparel from a good bunch of poultry. If plenty of range can be had, the cost of feed is comparatively nothing, and all the income a profit.

There is money in a bunch of scrubs, or a mixed lot of chickens, but there is more money in a good flock of pure bred fowls; they will lay more eggs, and bring higher prices on the market, they thrive better and grow larger, and besides all this, how much prettier a flock of chickens all of one variety look than a bunch of mixed up birds, and how much more satisfactory they are. A great deal depends on the start.

Do you want to get started right in this business? Write us your wants, and if we can't do it there is no need of your trying any one else.

Let us prove to you that there is money in eggs and poultry. Our prices are right and our goods are first class. Others are making good money in this business by obtaining their stock from us. Read some of the testimonials of our customers in this book. We satisfy them, we can satisfy you.

BEST TIME TO GO INTO THE POULTRY BUSINESS.

There is no time like the present to go into the poultry business. Just the sooner you get into it the better. Hard times will not affect it like other lines. There is always a good demand for poultry products and at a profit. The prices may vary a little and go up and down like all commodities; but it never gets so low that it is not profitable. So if you are contemplating entering the poultry business some time, act now and make the start at once. Remember now is the acceptable time.

SIX HUNDRED MILLION DOLLARS.

That sounds pretty big, don't it? Well, that is the size of the pile added to the nation's wealth by the poultry business last year.

Have you a part of it to jingle in your pockets, or for pin money? If not, then you are missing a good thing.

The poultry business offers better chances for more people of all classes and all conditions every place than any other line.

The poultry business cannot be overdone.

There may be a good many mistakes made in starting, but the conservative person who starts right and then goes ahead and plans his work along right lines and proceeds with care, knowledge and judgment, will win out as thousands and thousands are doing.

I can start you right and a chance at that six hundred million dollars.

WHO CAN ENTER THE POULTRY BUSINESS?

There is no other business that will admit such a number of people, with a full assurance of success as that of poultry. They can come from all the walks of life, all vocations, all classes, all peoples and make the poultry business a profitable one.

As it is a very healthy business, much of the work being out of doors and the work not hard, thousands of broken down business men, clerks, worn out traveling men, railroad men, miners, tradesmen, machinists and professional men who are eking out a scanty existence in overcrowded professions are turning to this industry.

Thousands of maidenly inclined school teachers, shop ladies, dressmakers, factory and sweat shop women, lady clerks with overstrained nervous systems and their vocations becoming a burden to them and the profits only making them a scant living, could take up the raising of poultry with great advantages for bettering their conditions. In the work of poultry raising, they would be gaining in health, obtaining a good living and laying up something for the proverbial rainy day. A peculiarity of the poultry business is that it leads to great esthusiasm and it is very easy to acquire a great liking for it. Read such valuable books as this "Profitable Poultry." Subscribe for a few of the many poultry journals and you will readily succumb to an attack of the "chicken fever." It becomes very violent sometimes and we have seen persons who could not think or talk of anything but chickens. It shows that there is a great fascination for the work.

I wish to cite you one instance, although I could point out thousands of them. One right here in our own state. Mrs. D. C. Johnson, of Maxwell, Ia., left a widow at 24 years of age with three small children and practically no means of support; education limited and with no business knowledge, started in the chicken business and in a few years, by sheer force of grit, push and hard earned ability, learned the business, made a living, made money and in five years bought a 270-acre farm with a good payment down and in a few years later had it paid for. What she and thousands of others are doing you can do. Here is what she says: "Poultry raising if started right and persisted in, can not fail to yield good returns in cash, health, strength and all that goes to make life worth living."

POULTRY BUSINESS OPEN TO ALL.

What would be the natural consequence of lessening the industrial and manufacturing business of the country? An overproduction of labor in all lines, which will lower the price and also throw many out of employment.

The poultry farmer is always sure of a good living; it will affect them much less than any other class. The farmers furnish the food for all classes, and as all must eat their products, which will always be in demand, perhaps not quite so high in hard times as in prosperous periods, but they will be the last to be affected. The farmers as a class have been making more money than any other.

The poultry raiser is strictly in it at all times; the supply is never too great.

The United States never produces enough eggs for the demand and imports great quantities from Canada every year. So there is no danger of overdoing the poultry business, especially in the production of eggs.

There is no business that affords greater opportunities to all classes than that of the poultry business. Take the great West, for instance, where many places in the new states and territories, eggs sell for from 30 to 60 cents per dozen. There is no business that can be started with so little capital as that of poultry, and it is open to all.

It is a business that is easily learned and readily picked up by any one with common intelligence. It does not require long apprenticeship or years of study. Just go at it and not too fast at first, and you cannot help but succeed.

There are opportunities for raising poultry everywhere; better near a town or city where a good steady market can be had for good poultry and fresh eggs. Young men just starting out in business, investigate this great enterprise. Small or unsuccessful farmer, look into this matter. Discouraged tradesman and laid off workman, it will pay you to investigate the great possibilities of the poultry business. It will put new life in you—new hope, and make life a thing of joy and happiness. It is open to all classes and conditions of people everywhere.

POULTRY BUSINESS FOR PHYSICALLY DISABLED PEOPLE.

On account of the poultry business being light work it is especially adapted to persons that are physically unable to do hard manual labor or perform hard, strenuous mental strains as the active business man is forced to do.

I do not want to lead you to think that the poultry business is all play and requires little or no work with hands or brain, but will say that it requires very close watching of small details and constant work—not hard work but constant light duties that because of their interest and fascination become a pleasure and the change usually has a very beneficial effect upon most people that are broken down in health and take up the chicken business.

And especially it is a beneficial change for those suffering mental disability, nervousness and the change of life of women, as the chicken business is highly fascinating, so much so that the "Chicken Fever" is very easy to catch and is one of the best things a person can have. It is good for health, pleasure and profit; you forget your ailments in the fascination of the work.

I think that it should be called the "Chicken Fancy" instead of "Fever," as the term is now applied to the thing or spell that comes over thousands of people and makes them talk chicken, think, plan and act chicken, even dream of chickens, of fine birds they will raise and the prizes they will win, and above all, the money they will make.

It is not all a "pipe dream," but a reality, as you will see by the thousands that must be making money where you concede the enormous amounts of this product that are marketed every day.

CUSTOMER SURPRISED AT SPLENDID CONDITION OF CHICKENS AFTER SUCH A LONG JOURNEY.

Pittsburg, Pa.

Dear Madam:—We received our chickens today and I was surprised at them being in such good condition as they were in such a long trip. I will also say that I am more than pleased with them.

Yours truly,

D. N. JOHNSON.

What has been Accomplished in the Poultry Business

Some Actual Results. You Can Do As Well.—What a Woman Did With Poultry.

Mr. and Mrs. Martin, of Dallas, Texas, started a few years ago on a small lot. Mr. Martin worked in Dallas. Their city home was practically paid for with their first products of poultry. They decided to buy a little farm, so chose one three miles from the city limits. They kept White Leghorns and White Rocks and made good progress in building up a beautiful home and successful poultry plant.

To quote Mrs. Martin's own words, "You see in 1910 we made a clean profit of $2,500.00 above all expenses, and being a good way off of the car line, bought an automobile to make quick trips to town and it proved a profitable investment."

When they first went to the country they had to do without the luxuries they had been accustomed to in the city, but today their home is equipped with every convenience of the city home, electric light in house and barn, telephone service, sewerage, bath, their own water system. The water is piped from a 360 barrel tank to various parts of the farm. Tank is kept full by a wind mill and gasoline engine. The tank tower has been walled in and forms the neatest kind of engine room. On the ground floor are mills which grind all the corn, beef scraps, and bone. These are driven by belts run to the second floor and drives a small saw mill; there all the lumber for shipping and show crates are prepared by Mr. Martin, who now spends his forenoon at the farm.

Mrs. Martin keeps the books and handles all the correspondence, besides mating and selecting all birds to be shipped or bred.

So to the woman who is puzzling over the problem of spending money or even of the necessities of life, why waver with indecision between many ideas, as poultry raising offers a pleasant and remunerative occupation. Opportunities for a repetition of Mrs. Martin's success are open to all.

HOW A MAINE POULTRYMAN WON SUCCESS.

Mr. O. D. Wells, of Skowhegan, Maine, has proved that a man with no previous experience may enter the poultry business and make money, if he goes at it right. Up to four years ago Mr. Wells had been engaged in the canning business exclusively, but when a tempting offer came for his plant and good will, he promptly accepted it. Then he embarked upon his venture with poultry and made good at once. The first year's receipts exceeded the running expenses by approximately nine hundred dollars, the next year by one thousand dollars, and the year following nearly thirteen hundred dollars. This sum represents compensation for his labor with the interest and the depreciation on his investment. He values the entire property at $4,200, including a fine dwelling house.

Mr. Wells started with the best he could procure and keeps improving them.

About 250 yearlings are carried over each year to be used for breeding. New blood is introduced through purchased males every generation. Mr. Wells does not take much stock in the trap nest theory of breeding, but prefers to depend on lots of vigor. He says that if a pullet has plenty of vigor she can't help laying if given the stuff to make eggs with.

To get at exact averages of cost and production it takes more bookkeeping than Mr. Wells can find time for, but he thinks that the feed costs about $1.50 per hen, while the corresponding income has never fallen below $3.00.

A PROFIT OF $5.00 PER HEN.

At Essex, Iowa, a town not far from Clarinda, in this county, our friend A. D. Murphy, who is a breeder of Barred Plymouth Rocks, made the statement that 60 hens brought him $300.00 worth of eggs last year, about $5.00 profit per hen. He makes good money with his poultry and is very successful.

A GOOD THING SOMETIMES TO BE THROWN OUT OF WORK.

Take the case of W. J. Tilley, Eastern, Conn. For years Mr. Tilley was a mill employee. Then, one day, when the mills shut down he had to find some other way of earning a livelihood. He had kept a few hens on the side, and decided to risk all that he had in a poultry farm. Now, he has a nice plant, with a capacity of a thousand hens; and he makes $1,500.00 or more a year; he is glad the mill shut down when it did.

NOTICE THE RESULT OF BIDDY'S HAPPY CACKLE FOR MR. CHRISTIE.

Mr. F. W. Christie, of Stony Point, N. Y., was a suburbanite when he succumbed to the seductions of biddy's happy cackle. Now, he has a modern plant on the side of the rugged 300 foot hill. He has five long houses for his White and Brown Leghorns and an incubator that will accommodate 6,000 eggs at one time. He runs his business of producing eggs just as a manufacturer would run his mill.

PANIC HAS NO EFFECT ON CHICKEN BUSINESS.

Henry Dana Smith was a business man in Rockland, Mass., when the panic caught him a few years ago. He quit the city and started raising poultry on a scrubby farm in a neighboring town. Soon, he began to specialize in roaster chickens and now he is known far and wide as one of the most successful poultrymen in New England. He raises five or six thousand chickens each year, and has developed a beautiful rural home, with a view for many miles across the country toward the sea. Mr. Smith does not try to work the farm. He raises fancy chickens and nothing else.

POULTRY BUSINESS NOT ONLY PROVES PROFITABLE, BUT RESTORES ONE TO HEALTH.

George A. Cosgrove was past middle age and broken in health when he left his Brooklyn flat to look for a farm in New England. The place he bought was a long drive from the station in Norther, Conn. For an untrained city man to attempt making a living in this semi-isolated spot, far from the markets, seemed like a desperate undertaking. Poultry was made the mainstay of the farm and within ten years the value had been doubled, a bank account started and health improved, likewise both Mr. and Mrs. Cosgrove have found contentment and peace of mind such as they had never known while living in the city. Of course, they were temperamentally adapted to country life. Some people would perish of ennui if they had to spend a month so far from the lights of the great white way. It ought to be said, too, that Mr. Cosgrove found five hundred hens ample to yield a satisfactory income.

This man makes the poultry business pay. Doesn't it look prosperous? See his incubator.

LARGE INCOME FROM POULTRY BY THE LONE EFFORTS OF A WOMAN.

Mrs. Andrew Brooks, who lives in the Northern part of New York state, is doing a $1,200.00 or $1,500.00 business by her own efforts alone. She is a modest but enterprising little lady, and knows poultry from A to Izzard, so to speak; or, as one might say, perhaps, from feathers to gizzard.

TWO MORE SUCCESSFUL WOMEN.

The name of Mrs. Basley has long been a familiar one among poultry raisers of the Pacific coast, especially those interested in White Plymouth Rocks, her favorite breed. Hundreds of women who do not attempt to make a living by the raising of poultry are still adding largely to the family income by the means of the eggs they sell. One New Jersey woman was able to lift the mortgage through the aid of the hens that her husband had good humoredly joked about.

DEVELOPMENT OF A 5,000 CHICK FARM.

Henry D. Smith was engaged in the manufacturing business when fire destroyed his plant and left him almost stranded. However, he managed to get hold of a farm by the aid of friends and started in the chicken business. He made good from the start.

Together with his two boys, one 13 years of age, the other 15, and his faithful wife, to whom much of the credit is justly due, they have developed a wonderful profitable business in chickens, raising over 5,000 last year, and 1,800 squabs. He works principally for broilers and early frys.

He is now glad that fate, which seemed to be so cruel, started him in the chicken business, as now he has a plant that is a better money producer than he could ever expect in his manufacturing business, and it is so much more satisfactory. Away from the smoky, crowded city to the care-free, clean country. The only life to live.

WHY THOROUGHBRED POULTRY PAYS.

The American Hen in one year earns enough to buy all the gold and silver dug out of our mines, all the sheep in the country and their wool; it is also said that there would then be a balance left to buy all the buckwheat, rye and potatoes raised in the U. S., and even after all this there is a vast sum left that we could not comprehend. When it is figured down in this way and itemized or analyzed in such a way that we can understand what a vast sum really is and should be credited to the poultry we begin to wonder whether or not we have our share, and if not, why not? It also shows that some one is reaping benefits and there are possibilities for every one. Read in another article why it is the thoroughbreds that will swell the already large returns and what a great advantage the purebreds have over the mongrels. We now have the problem where it is easily solved. Mongrels with the few thoroughbreds in past years have made the Nation what it is today; what will all thoroughbreds in the years to come mean to the U. S? How many men will get in on the ground floor?

"Chickie, Did You Lay This?"

GREAT ADVANTAGE OF THOROUGHBREDS OVER MONGRELS.

I know that the poultry raiser is not very much interested in poultry statistics and calculations, but a few reports will not go bad and tend to prove what the prospects in a certain enterprise are. It matters little what vocation is attempted, it is first accomplished by an investigation. Now, let us turn our attention for a few moments to the practical side of the poultry question, having to do principally with what the large packers and poultry buyers, those on whom we depend for the exchanging of our poultry into medium of exchange called "MONEY."

I will say right on the start that if a poultry raiser replaces mongrel stock with thoroughbreds he will receive from 25% to 100% more for his poultry and eggs, with no increase in their cost of maintenance. This fact should interest every man whether he keeps five chickens or five hundred.

A few questions have been put to the packers and buyers who purchase millions of pounds of poultry yearly and their answers should verify beyond the question of a doubt that thoroughbred stock will bring better prices than mongrels. Some of the questions and answers are as follows:

Do you pay a better price for thoroughbreds than for mongrels, and is the discrimination in price because they weigh more, or because they are more desirable for market purposes?

The answer was that a better price was paid for the thoroughbreds, because the yield of meat is greater with less bone. They sell for more at retail, are more shapely, more uniform, in fact in every way more desirable for market purposes. This is a pretty strong evidence in favor of purebred chickens, especially as coming from men who practically supply the civilized world with poultry products.

Another question was asked, if any effort had been made by the packers to stimulate an interest in the cause of thoroughbred poultry, and they were also asked what per cent of the poultry bought was thoroughbred. To these questions the reply was that some effort had been made to stimulate an interest, but not to any extent. They claim that only one-fourth of the stock they buy today is thoroughbred (some of them, however, will buy nothing else). The average price ranges from 8 to 16 cents per lb., and the average weight is four and one-half pounds. All the leading packers agreed that thoroughbreds would weigh more than mongrels on similar feed rations. Some estimated their weight at 75% more.

The reason that a poultry raiser should buy thoroughbreds for his parent stock is simply because superior qualities can be obtained in no other way. It it not altogether because the feathers are white, black or buff, or whatever color might be tasteful, or because the ancestors of said flock had some great winning records, but the principal factor back of the selection and choosing of the thoroughbreds is the fact of their money-making values, big dividends on the investment, sums it all up nicely.

TRUTH IS GOLDEN.

That is much more than a proverb, my friends, it is a real actual fact to be dealt with in every transaction in life.

If the truth was only fully known in regard to the enormous and wonderful profits in the poultry business, there would be many more people bettering their conditions by entering the business. There is ample room for thousands more to enter this profitable business and there would certainly be a grand rush if the "truth" was only known.

When you know the truth, the actual unvarnished facts in regard to poultry, and how to get a safe and easy start, it will be dollars and cents in your pockets and would indeed be golden—actual gold of the realm.

And then it is golden to know the truth and unvarnished facts and information about poultry as I have given it to you here in this book. The following of the Golden Rule, as is practiced on our poultry farm, is surely Truth Golden, and actually enters into every transaction we make.

GREAT ADVANCE IN PRICE OF POULTRY PRODUCTS IN THE PAST 10 YEARS.

The value of eggs produced last year averaged 20 cents per doz. at the farm or poultry yards, this for the whole U. S. compared to only 11 cents per doz. 10 years ago, according to the federal census.

Hence the value of eggs produced last year was three hundred and forty-two million dollars compared to one hundred and forty-four million 10 years ago, so you will observe that values have a good deal more than doubled; the extraordinary importance of this factor is further evidenced by these facts: The gross value of poultry last year was $1.80 for each fowl on hands at the close of the year, compared to $1.10 ten years ago. For each dollar's worth of poultry on hands for this time, the gross returns last year exceeded four dollars, compared to but a trifle over three dollars ten years ago.

An egg is a little thing, but it is not considered too unimportant for consideration by the U. S. Department of Agriculture. In view, however, of the fact that it was shown by the census of 1900 that the yearly production of eggs had reached 1,293,662,433 dozen, and the value of poultry production, including eggs, was in that year $250,623,114, this interest is not surprising, especially since we are advised that the value of poultry and eggs is increasing in the State of

Kansas alone at the rate of $1,000,000 per year. In the production of eggs, Iowa leads all other states, and the agricultural department has published a table showing the profits on the eggs in that state from the time they leave the hands of the farmer until they are placed in the basket of the consumer.

This is certainly gratifying and shows the results of purebred poultry, and great attention being paid to the breed advancing to the standard in egg production as well as size, quality, hardiness and vigor. The argument is certainly strong that it pays to get good stock and give them good care. It also shows that there is positively great profit in the poultry business. Great is the hen.

BETTER LATE THAN NEVER.

It is strange, but nevertheless a fact, that people will spend the best part of their lives in hard work, becoming broken down, their health becomes poor, and very little laid by in the way of money for the proverbial rainy day, so they are forced to look for some other means of subsistence other than a trade or occupation, that they had spent the best days of their life and only receive a living.

Many turn their attention to poultry, and I am glad to say that in a great many cases they make good, even when broken in spirits, health and financially. They can enter the poultry business and take on a new lease of life, recuperate their health, renew their old time spirit and not only make a good comfortable living, but generally succeed in laying by a good sum to keep them in the ev'n time of life, while often they receive a great many of the luxuries of life as they go along. The question is, what would they have accomplished had they started when they were in their prime, with all the vim and energy of youth and the best days of their life? This is a suggestion to the young man looking for something that is profitable and pleasant.

I am glad that the poultry business is so admirably adapted to that class of people and that it enables them to commence this very fascinating and profitable occupation, later in life after they have met with the vicissitudes, reverses, and traveled over many of the rough places, so it is better late than never, to go into such a profitable business as the culture of pure-bred poultry.

THE GRAIN THAT MAKES THE "STAFF OF LIFE" OVERBALANCED BY THE POULTRY BUSINESS.

HOW TO MAKE MONEY AT HOME.

This is a letter written by a bright young lady friend who tells her secret. Good advice, and shows a success. You can do likewise.

Young folks, why leave your farm home because you want to make money? Why not stay on the farm and make the money there?

I speak from experience—and this is my secret—go into the poultry business, but go into it right. Look the problem squarely in the face, make up your mind to spend enough to start right, and to read, study and work enough to keep right. That is what I did and my hens sent me to the St. Louis exposition with $100.00 to spend on my trip.

My first step was to put up a comfortable house—lined throughout with tarred paper. I got the cheapest mixture of paint I could find and painted the paper, roosts and every crack and crevice, stopping every possible chink in which lice might gather. I had long, rough sheds put up for hens to exercise and scratch in, in bad weather. The breeds I selected were Wyandottes and Plymouth Rocks. Both these breeds are prolific layers. I got a small feed chopper and a big iron pot. For evening meal I always gave my hens a good hot mash of first one thing and then another. Sometimes it would be potatoes boiled with scraps of meat, parsnips, etc., from the kitchen, and then mashed with meal or bran put in to thicken (never feed sloppy food). Then again turnips, cabbage or beets would take the place of the potatoes. Several times a week I would chop a lot of clover hay and put it in to scald well. It would come from the pot as green and nice as in summer and was a special favorite with the hens. In the morning I gave a heavy feed of corn, wheat and some grain.

Never let your fowls suffer for plenty of fresh, clean water or variety of food. Twice a week in winter I would go to the butchers and for ten cents get a great pan of bones and scrappy meat. This would go through the crusher and furnished the meat and lime necessary to keep poultry in a strong, healthy condition. I have even given as high as $1.15 a bushel for wheat for my hens and they amply

The Grain That Makes the Staff of Life Overbalanced by the Poultry Business.

repaid the outlay by the increase in eggs. I raise a lot of red peppers in the summer and always use them freely in the morning mash, as it helps keep poultry warm and active. Send the hens to roost with a warm craw.

In the scratching sheds was always plenty of broken up hay, straw, etc., into which I threw the grain, thus making them scratch and work for their living as in summer.

To prevent lice I got a fifty-cent sprayer and every now and then on warm, bright days would catch each hen and spray her well with gasoline. It would make them dizzy for a little while, but did them no harm and was a sure preventive. Each time I used it I would spray a little around the hen house and in the nest boxes.

I did not let the home merchants have my eggs, but made arrangements with a hotel in my nearest city to furnish so many hundred eggs twice a week at so much per dozen. I guaranteed them to be clean, and perfectly fresh, and to prove that I knew they were the best to be had I stamped my name upon every egg I sent off. I never kept a hen after she was two years old and the hotel manager was always glad to get what hens I had to sell and gave good prices for them.

Of course, I admit that there is a lot of work, and some of it very unpleasant, but soon you will learn to love your flock of biddies, and enjoy feeding and attending to them, and what a delight it is each evening to take your basket and go around to gather up the eggs. Won't you take my advice and make the effort?
—Miss McVeigh, Culpepper Co., Va.

HOW TO MAKE POULTRY RAISING PROFITABLE.

This is a paper read before our Page County Farmers' Institute ten years ago. It seems to me that it is good enough to put in this book.

GOOD STOCK.

BY MRS. A. A. BERRY.

Get a good breed of fowls. Almost any of the standard varieties can be made profitable, and it seems that almost every chicken fancier has his choice of breeds for various reasons that only he himself can give. For general purpose fowls on the farm, select one of the large breeds. If eggs are the only object the smaller breeds are generally better, but the farmer desires a fowl that will produce eggs, be suitable for the market, and when prepared for his own table be large enough to satisfy an abnormal appetite of a medium sized family. It pays no better to keep "scrubs" and mongrels in your poultry yards than it does to keep "scrub" horses, cattle or hogs in your barns and feed lots. The "scrubs" in all lines are back numbers. So we advise you to get out of the old rut and join the ever increasing procession of raising good fowls. It costs as much in time, labor and feed to raise a mongrel that weighs three pounds as a Plymouth Rock or Cochin that weighs six pounds at the same age. Get thorughbreds, if possible; if not, get good grades. Never be satisfied with anything but thoroughbred males for breeding purposes. Your flock of hens will be better separated from the cockerels except during the breeding season when you are mating for eggs for hatching purposes. We repeat that the very foundation of success in poultry, as in other lines of industry, is good stock. This is the first requisite. See that you have it.

GOOD SHELTER.

Cows will not produce large quantities of milk sheltered on the sunny side of a barb wire fence, when the mercury is hovering around zero. Neither will a hen lay a hatful of eggs every day if she has to roost on the limb of some tree or out on a self binder that cost you $125.00. In short, have good shelter—not an elaborate or artistic affair supplied with artificial heat and ventilation, but a good, substantial plain hen house. It should be boarded up with drop siding, or if with plain boards, the cracks must be covered with battens to keep out the cold wind. It is also a good idea to have it lined with heavy building paper and banked up around the bottom with stable manure or old straw. Stop all the cracks where cold wind can get through.

A good start in the poultry business, as little Mable's mamma follows the teachings of "Profitable Poultry" and is a good customer.

Don't worry much about the ventilation in winter time. Very few chicken houses are built so tight that they require extra ventilation. There are more that have too much than not enough. Above all things avoid dampness and draughts. Nine times out of ten they are the source of roup, colds, etc.

GOOD CARE.

Use good common sense in caring for your fowls. Give them the same care and attention that you do your horses and cattle and then you may expect results. On many farms the flock of hens get very little care, perhaps a little whole corn thrown to them on stormy days and no shelter except what they are able to find for themselves about the sheds and farm buildings. Other stock about the place would perish with the same treatment. Then people wonder why their hens do not lay in winter when eggs are scarce and a good price, and because they do not, they conclude that poultry keeping is unprofitable.

PROPER FOOD.

When the dairyman wants to produce milk and butter, he accordingly feeds all the good, wholesome milk producing food the cow will readily consume. When the farmer wants to produce pork, he feeds them more than simply enough to keep them in ordinary flesh; he feeds fat producing foods. No one will dispute that this pays. It has proven a fact too often. Statistics show that the value of the poultry products of the United States last year was more than the value of the wheat raised, and more than the combined value of all the sheep and hogs produced. In the face of these facts it pays to give the hen a little attention and study her and her requirements.

The reason hens lay in the spring and summer is because they are able to find egg producing elements in part. By making the conditions the same in the winter, she will lay. She can't help herself. She has no control over the egg production.

Now, what are the summer conditions. Warmth, grain and animal food, found in bugs, worms, etc., also green vegetables. We have given some hints for producing the first essential, warmth, and now we shall say something about the feeding. In the evening, always give a warm mash. This should be composed of ground grains, wheat, bran and animal matter in the form of blood meal, meat meal and meat scraps.

At noon, feed whole wheat, wheat screenings, kaffir corn, cane and a small quantity of oats. It is a good plan to keep a portion of this grain in the sheaf. Make the fowl hustle for all their feed, it keeps up circulation. The morning ration should compose of little mash, plenty of grain fed as above suggested. Never, under any circumstances, allow feed to accumulate about the feeding pens. If for any reason they do not clean up all the feed given them, cut down the quantity. Be sure to give them just what they will clean up—no more, no less.

In the winter, especially during the extreme cold weather, they should be fed in the house and not allowed to run out in the yard early in the morning, as they are much more liable to become thoroughly chilled, and their combs and wattles frozen, by getting out too early in the mornings, than they are during the night. Give them their morning mash inside, and allow them to stay there until the sun has a chance to warm up the outside atmosphere a little.

Every regulated hen house should be provided with a small yard, part of a shed if possible, and located on the sunny side of the building. Every few days a quantity of fresh straw, chaff, leaves or other littler should be scattered over this pen and all the grain rations should be scattered among this litter. By so doing the fowls will have to scratch among the litter to obtain their food. It does a hen good to "Scratch for a living," if she is sure to get something where she scratches. As in all animal life, exercise is one of the requisites to the health and highest development of the hen. A hen that does not exercise becomes lazy, too fat, and non-egg producer, under which conditions she is fit only for the hatchet.

In addition to the foregoing suggestions for feeding poultry for eggs, provide some kind of vegetable food, as near the green state as possible, such as cut clover, alfalfa, etc. Also feed cabbage, turnips, small or refuse potatoes, table scraps or almost anything that they will eat and that will furnish a variety.

It is an excellent plan to tie a head of cabbage, beet or turnip up on a wall, just low enough for them to reach with a little effort. This will furnish exercise which will be good for them, especially on stormy days when they have to stay in the house.

A large amount of green food can be procured by raising a small patch of rape or kale. These plants belong to the cabbage family and can be stored in the same way. If fowls are kept in confinement, a small pasture should be provided, sown to foliage producing plants. Provide plenty of grit in the form of coarse sand, bone meal and oyster shell. Grit of some kind should be kept within their reach at all times.

Provide plenty of good clear water. Do not, under any circumstances, allow the drinking water to become foul and unfit for use. The drinking vessels should be cleaned every day. In cold weather give warm water. Your fowls will appreciate it and be less likely to have their combs and wattles frozen. But with these suggestions use common sense and good judgment and you are bound to succeed.

LITTLE CAPITAL REQUIRED.

In many ways it is a fact that the less money a person has in starting the chicken business, the better off he is and the better is his chance of success;

as the stern necessities seem to be the only power that can induce a person with the poultry fever to let their plants grow naturally. The chicken fever often befuddles his brain and the crazy notion that he must start big and at the top must be gotten rid of. Start in a small way with good pure birds if you do not know much about the business, and grow naturally. It is is not capital that is of the most importance to the poultryman; it is knowledge of his business and actual experience. There are plants costing many thousands of dollars. We know of one plan that is raising poultry and eggs for market that is earning $20,000.00 per year profit. But it started in a small way, as all the large and successful plants have done.

As to the exact amount of capital required, it all depends upon your circumstances, and we could safely say you could start in the poultry business with from $10.00 to $5,000.00. The latter sum could well be paid for a twenty or forty acre farm, well improved, and near some large town. If land and property is high, one may go out into the country a little ways and meet the requirements of his pocketbook. The cheap, rough land makes the very best poultry farms, so one can always find the place to meet their requirements if they have the will to do so.

As to cost of starting, one or two good incubators, one or two good brooders and a small flock of pure bred chickens of one breed to start with, then you can add more as you increase.

WHERE TO START IN THE POULTRY BUSINESS.

We receive many letters asking my advice as to where would be the most desirable place to start the poultry business.

There are thousands of desirable places that most any one could make it go to start a poultry business. The nearer you are to a good market the better, although there is no other product that there is such a universal good market everywhere and in such demand than poultry and eggs, and especially eggs; but of course there are places where prices are higher than others. There are cities of five thousand and upwards in the states of Montana, Wyoming, Utah, Nevada, New Mexico, Colorado, California and sunny Florida, that the average prices of eggs the past season has been from 40 to 50 cents per dozen and the price of feed not unreasonable.

Land not too high and improvements in the way of buildings, fences, etc., are moderate.

There are also great opportunities in the crowded east with large cities and a wealthy class, which find it hard to get a strictly high grade of fresh eggs and fine, juicy, tender fries, roasters and broilers. They are willing to pay a nice premium to get the best, such as is easily raised and produced as what is now put on the market. You can obtain a line of customers that you can contract with to furnish a stated amount the year around. No trouble in getting customers and no trouble in producing the goods that will fill the bill if you know how, and as many others are doing it, you can too.

California, especially around Pataluma, is a great place for egg ranches. There are more chickens raised for market and egg production within the radius of 50 miles around Clarinda than any other spot in the world. Census proves this statement.

There are hundreds of small poultry plants and in fact nearly every farmer raises from 200 to 1,000 chickens. They find that it pays or they would not keep it up.

A Fine Flock of Toulouse Geese.

If you are looking for a location to go into the poultry business, look around in your own vicinity, you can generally find an opening. If you cannot decide, write me, giving conditions as to location, prices paid for produce, etc.

I will help you to locate and if you wish a change of climate for your health, I can cite you to lots of openings in the West, South and North; there are many desirable places in every state. I am especially interested in "Sunny Florida." I know of some splendid openings to combine the chicken business with fruit and vegetables, so write me and I will give you such information as I am capable.

GET STARTED RIGHT.

Every one wants to make a success of whatever they undertake. All do not have the same object in view when they go into the chicken business; most people go into it for money or a living, many go into it for diversion, a change, something to occupy their mind or strictly for their health; but of course all expect profit. No matter what your object, it is highly important to start, not necessarily expensive, but with the right kind of stock, eggs and appliances and at right prices.

You could squander a great deal in starting and then not get what is best suited for your requirements. I have made a study of the poultry business in all its details and especially in starting beginners, so I can be of great assistance in starting.

I do not consider that my part of the transaction is carried out when I fill your order after receiving your remittance. I must know that you make a success of it or help find out the reason and point the remedy, but I tell you, my interested friends, that the start is the one great important factor in making a success. I can start you right by furnishing such stock, eggs, incubators or supplies that are bound to lead on to success no matter what your object of entering the chicken business is, if you do your part. I can not only start you right and see that you do make a success of it, but assist you in keeping right and doing the proper things at the proper time, as thousands of my customers are.

POULTRY BUSINESS AS A LAST RESORT.

Too often it is the case that people will enter the poultry business only after having failed in everything else they undertook, and then when their best producing days were over took up the chicken business as a last resort. Of course it is "better late than never," and generally they can make good; but it is a great deal easier to make a success of it if started earlier in life or before the best producing days were over.

The chicken business is a grand opportunity open to all, but you cannot expect to make the big success of it, as if you had gone into it with your best, instead of as a last resort.

Many think that when they get too old to make a full hand at manual labor that they can take up the chicken business. They can and generally will succeed fairly well—at least make a good living; but it would come much easier if they had saved up a tidy sum, enough to buy a little farm or suburban property and equip it with buildings and stock it properly, starting with good pure bred poultry; many start with just a few eggs, an incubator is also necessary to make a success of the business. I can think of no other business you could start with such a small amount of capital.

I want to impress upon you that the great advantage in starting the chicken business is to begin early in life—the "Hay-day" of life, as the saying is. You can start the business and keep up your other work and learn the business thoroughly. Eight and ten hours in your regular work leaves you plenty of time to raise lots of chickens, even if you are only making a living at your regular work or vocation, you can start the poultry business as a side line and learn the best of all occupations, which in time will make you successful and independent.

There is hope for even the man "down and out," who has made one failure after another proving that he has the right kind of stuff in him.

I have given all the information in the book necessary to make the start and carry out your plans, but you should have a little capital, not necessary to have much, or try to start big, the idea is to work the business up to a paying investment, start with only a trio of fowls, or a small pen, some get a start with eggs, all you need is the foundation and you need not spend more than you can afford, yet success is assured. As I have already stated, you can make the poultry business a side line, until you get it on a paying basis, thus doing your regular work to help along in building the poultry business on a sound, substantial basis.

Young men—young women, I urge you to investigate the great possibilities of the chicken business and do not take chances of everything else before you try poultry. The opportunities of this pleasant, profitable and fascinating business were never so great as they are today and the future was never so rosy and bright. The opportunities are marvelous. Will you grasp it?

HOW TO START THE CHICKEN BUSINESS.

A Fine Brood.

This is a question that is asked us a great many times in the course of a season.

There are a great many things to take into consideration in deciding the best course to pursue in starting the chicken business. If you have no experience at all and have only a lot or two, just buy a pair, trio or pen of general-purpose fowls; some family of the Rocks, Wyandottes, Reds, Orpingtons or a heavier breed, such as the Cochins or Brahmas. The variety is a matter of choice as to color, size and form. You will not go wrong to select any of the above varieties.

It is not necessary to get expensive buildings and appliances unless you have plenty of means at your disposal. If you have some experience and wish to enlarge or go into it for a living for profit, buy or rent a place either small or large and as near a large town or a good market as it is possible, combine the business for furnishing pure-bred birds and eggs to your neighbors and selling the surplus on the market. You can sell lots of breeders and eggs for hatching right from the start, but it is a business that takes several years to get your name advertised. By obtaining stock and eggs from me gives you a good advertisement to start with, but you have to work up your own reputation as a breeder and learn the art of mating. I would help you do this through correspondence.

It even takes a little time to work up a reputation to furnish absolutely fresh eggs and fat, juicy birds, but it is not hard to do if you go at it properly.

Land does not need to be expensive, as poor, thin, rolling and sandy soil is the best for chickens.

If you have several acres to devote to the chicken business, get a good laying variety such as some family of the Leghorns, Anconas, Andalusians, Minorcas or Hamburgs, and also a good general purpose fowl, these two will be sufficient, don't try too many varieties until you get started well and then you will know by experience whether or not you can handle more successfully.

If you are in the city and want to go to the suburb or country and you can't find a place that suits you or wish some advice, write me; I often know of good openings. Tell me your circumstances as to how much you have to start with and your experience. I find lots of people in town or even in small country villages who don't know a great deal about the real poultry business.

Just look around, ask questions. Don't let "Grouches" discourage you, there are too many "grouches" in the world. Write me and send for a few birds or buy some common stock of your neighbor and send for a few settings of eggs of the pure-bred stock or buy an incubator and enough eggs to fill it. Ask some more questions of a successful poultry fancier and write me again on any subject that has not been fully described or is not clear to you. You will ge there all right when the hen fever gets hold of you.

YOUR OPPORTUNITY.

If you knew that you would be a success in the poultry business you would no doubt start immediately.

You never will know just how successful you are going to be in anytihng until you try. What if every student contemplating law, medicine or a profession of any kind, thought the same way before starting to study. Of course they would fail. Did you ever hear of a person being a success if he were inclined to complain and look for excuses to keep from engaging in business.

Nothing can be accomplished without energy. The sooner you get started in the right track, the sooner you will have your little chicken business on the money making basis.

Do not lay this opportunity aside with a view of considering it later. Act Now.

YOU CAN WIN IN THE CHICKEN BUSINESS.

An opportunity to make more money honestly is now presented to you. Why work for others on a salary when you can do better. "He who hesitates is lost," is a well known maxim.

I am anxious to start you and willing to give you the benefit of my own experience in a way that should assist you to overcome every obstacle. The poultry business is peculiar in that there is a great amount of enthusiasm in it. Your whole soul will be in the work and with persistence and determination you will win out, I am sure.

You should not be foolish enough to believe some people that will persuade you to give up your present occupation and plunge into something you know nothing about, but take my advice, which I give to you frankly and fully, that the best thing to do is to remain in your present position if you have one until you get fairly well started. The business at first will not consume all your time, but of course as the business develops and your flocks increase, you will be compelled to devote more time to it.

MAKING GOOD.

In this day and age people will not accept all that they hear or read, but have to be shown, statements made have to be proven. Women, however, are not so imaginative as men, as my observation is that in business they are more serious and are more careful of their statements.

I have made a great many sweeping statements in regard to the advantages and profits of the chicken business, but please take notice I back every statement. I take statistics, which are surely good backing, in regard to figures. I give testimonials of judges and from my customers, from experimental station work and records and I cite you to thousands of successful poultry raisers and fanciers, and last but not least my own personal example of what I have done in the chicken business and the success I have achieved along that line.

I am only a person with ordinary intelligence starting at the bottom, started small and climbing up paying my own way as I went along, selecting, mating, breeding, culling, improving my flock, introducing new methods and advancing from step to step until I have a wonderful business in furnishing the finest of pure bred, heavy egg producing fowls, incubators and supplies.

If they were not the best for the money I could not keep my business up, whereas I double and triple it every year. I make good every assertion, not only because it is a pleasure to do so, but then I have to, as I put out the strongest Iron Clad Guarantee that any poultry concern of our kind puts out; every assertion is backed up by cold facts and an Iron Clad Guarantee.

MISTAKES TO AVOID.

The less mistakes you make the quicker you will succeed. One of the greatest mistakes many beginners make in the poultry business is to start with inferior birds, eggs, incubators and supplies.

First and above all things, get the best. Better pay a little more and get something you can depend upon, something that will lay you a strong foundation to start with. If you do not have a great deal of money to expend in starting up put enough in a pair of fowls or a setting of eggs that will produce good ones, such as we furnish at the Berry's Golden Rule Poultry Farm. While our prices are very reasonable and lower than most, the quality is of the highest.

It is a mistake for a beginner to start with too many varieties of fowls or eggs. It is all right to branch out when you are fully established, but it is certainly a mistaken idea for the beginner to order from five to ten different varieties at once.

It is a mistake not to start now. Why put off longer when you can get into the game NOW.

A WORD TO BEGINNERS.

A great many of our customers are beginners in the poultry business and I want to give a word of advice.

Don't build too lofty air castles. Don't expect too much, be rational and sensible.

I received a letter last spring like this, which is a good type of hundreds I receive every season:

Mrs. A. A. Berry,
Dear Madam:—I am starting in the chicken business, and this order I send you is my first. I have given up my position in a store and have leased ten acres for five years and have erected good buildings to start in the business.
I want to buy 200 hens and sufficient cocks and was thinking of starting with about five breeds. Which ones would you advise, etc.
Can I expect $1000 per year profit, etc., etc.

Now, the young man is perhaps like hundreds of others, and I take it is composed of the right kind of stuff to make a success of the business of his choice.

He made the mistake in quitting his job until he had learned the business and got the start.

Let us reason a little together. Suppose that this young man decided to be a lawyer, doctor, or took to some other profession, could he expect to learn the the business and start right off earning $1,000.00 per year? Or what business would you expect to engage in that would net you $1,000.00 on the start and with two or three hundred dollars capital? No legitimate business would be expected to do this.

Now, I would advise that young man to start by obtaining a pen or two of standard bred fowls of the very best which I could furnish. Secure a good yard or lot with a good house, coops, etc., and an incubator and brooder, and send to me for some standard bred eggs.

Keep your position and do this work mornings and evenings as a side line, or you can start it while finishing up high school, then when finished you can take a course in some business college during the slack chicken season, and in this way learn the business as you go along; you can then branch off and expand when you get the experience, as there are unlimited possibilities in the poultry business.

You must learn the business same as anything else; yet it is easier to learn than most any other business and I know it is more profitable.

I am sure that I do not want to disappoint you by selling you much more than you will need to start with. I want you to make the businesss a success instead of a failure, as if you succeed as thousands do you will be a frequent as well as a pleased customer.

There is perhaps no other line of business that is so generally successful on a small scale and a failure in a large way as poultry raising.

Big costly plants with flocks of thousands lose money through inexperience, while the same parties with a cheap outfit and limited flock show profit right from the start.

There is no avenue to success in the world today as broad and open to everybody as that leading to raising poultry.

But it is not carpeted with brussels or strewn with roses without thorns; you cannot expect to reach the goal on a big red automobile, you must expect to dig and keep going, same as you would in any other business of which you expected to make a success.

But I will say that there is a fascination, a pleasure in the work that few businesses possess.

That success is easier to attain than many other lines of business I know of, starting with the same capital; but it has to be learned by actual experience the same as I have done and all others.

I will be glad to assist you get a start and I would rather furnish you with a pen or two and a few hundred eggs than a $500.00 start; but if you have a start and somewhat familiar with the business and know the business and want $500.00 worth of fowls, eggs or incubators, I can furnish them, as I have done others. However, I give the same personal attention to an order for a setting of eggs or a pair of fowls as to a large order.

View of one of our White Leghorn Farms on Colony Plan.

EGG FARMING ON THE COLONY PLAN.

Poultry for egg production alone, is carried on extensively in many parts of our own country and England, some of them producing as high as $100.00 worth of eggs daily. Personally, we advocate this branch of the industry, and on what is known as the colony plan. These consist of a number of small houses described later, each of which will easily accommodate forty to fifty hens. They can be built in a large yard or field, if your premises are large enough so the birds will not wander away from home and bother the neighbors. They should be built fifty yards apart. If the hens are confined in a small yard beside their house a week or so until they get accustomed to their own place of nesting and roosting, there will be no trouble about their finding their own houses, or in their crowding some houses and leaving others almost entirely empty. This can be watched and if they get to straying in the wrong houses they may be divided. These houses will cost $10.00 to $15.00 each, depending almost entirely upon the kind and cost of material.

On the Golden Rule Poultry Farm we have a number of them which are 8x10 with brick foundation one foot deep built to keep out rats and other vermin that prey upon the young chicks. We batten the cracks and shingle the roofs and the cost is not over $18.00 each.

AS A SIDE LINE.

Poultry raising makes a splendid side line for any whose time may not all be taken up. It is surprising what may be accomplished along this line by those having an hour or so at their disposal each morning and evening. Many people who are otherwise engaged most of the day enter the poultry business in a small way for the pleasure it affords them; to say nothing of the satisfaction there is in having plenty of fresh eggs and tender poultry for use on their own tables. This phase of the business is especially adapted to those owning large city or town lots, although many make a success in a small way on one city or town lot.

POULTRY AND SPECIALTY VEGETABLE GROWING.

Poultry raising and fruit growing go well together, but it takes several years to get returns from fruit, except small fruit. Specialty vegetable growing is a splendid thing combined with poultry. We do not mean regular truck growing, running a vegetable wagon, but some special crop of vegetables, such as potatoes, cabbage, tomatoes, onions, or some vegetable that there is a ready market for and can be raised to advantage in your vicinity.

Potatoes are most plausible, as there is always a good demand for them everywhere and generally at a good price. You can obtain good returns from them.

If you are near a canning factory or are near a large town or city within easy shipping distance, tomatoes and sweet corn for canning, or cabbage, onions or some of the vegetables for market purposes are very profitable and can be combined with poultry.

It is true that many of the specialty truck growers, truck farmers, and regular farmers raise more or less poultry, but they could do much more at it, and many that do nothing could raise poultry with great sucess.

There is a great opportunity along this line for all. Choose a suitable location and go at it and you are bound to win, if you take off your coat and go at it in earnest.

There are thousands of opportunities, many openings right in your own locality.

Study this matter, make inquiry and see what can be done.

COMBINING POULTRY AND SMALL FRUITS.

These are the poultry yards of one of our friends who is a successful poultry and fruit raiser. Don't these hens look contented and as if they were ready to do their duty in shelling out the eggs?

These go together and there are many who are combining the poultry business with the raising of small fruits, with great success. Poultry is very profitable in orchards, even commercial orchards, besides doing a great good to the orchard. They can be raised cheaply, living on bugs and insects that are detrimental to the fruit. This is so with small fruit and the chickens do a great amount of good by keeping down the bugs, worms and insects that prey upon the fruit. They may be raised on the same land with only the additional expense of buildings, incubators, brooders and coops.

PURE BRED STOCK.

The very foundation of success to obtain profitable poultry is good stock. This is the first requirement; see that you have it. By all means, begin with pure bred poultry. It makes no difference what breed of fowls you undertake to raise or whether you are breeding for broilers, roasters or eggs, or for fancy stock, begin with thooughbred fowls. Buy a pair or two or use the eggs for hatching purposes or buy a full setting of eggs from a reliable breeder and set them. Money expended for stock either way, will be capital well invested. The advantages of pure bred poultry over scrub stock are great. Here we are ready to furnish you, in any quantity you may desire, and we will guarantee that they will please you and feel sure that they will be a splendid investment and you will never regret the few dollars spent for them.

IMPORTANCE OF THOROUGHBRED MALES.

Under no circumstances be satisfied with anything short of a thoroughbred male and that the best you can find or at least the very best your means will permit you to buy. Procure good, thoroughbred males every year to still further breed your stock up and keep them up to a high standard.

EGGS FROM PURE BRED STOCK.

A most excellent way and one greatly practiced to start in the poultry business is to procure some eggs from a good reliable breeder who makes a specialty of breeding good stock. This is the cheapest way, as you can get a few settings or enough to fill your incubator and then obtain the start for a fine flock of the breed of your choice. We wish to say in this connection that we offer pure bred eggs and poultry at a reasonable price and the quality cannot be beaten.

DO NOT LET ANY ONE INFLUENCE YOU AGAINST STARTING A GOOD BUSINESS.

I take it that you have about made up your mind to enter the poultry business, or enlarge upon the start you already have. You are convinced in your own mind that it is a good thing, a paying business. But here comes some well meaning friend or acquaintance who is a regular "doubting Thomas"—a fellow that is afraid of his own shadow and throws cold water on the proposition, and discourages you. Some people never get anywhere in life; just drift along, never make an important move, or start to better their condition. And they talk to their friends or whoever will listen, against doing something important, often so discouraging them to such an extent that they back out, after deciding to take an important step towards bettering their condition.

Don't be a pessimist, or listen to one. Be an optimist and have the backbone to carry out your own convictions, and get into the poultry business, because it is a good thing, no difference what any one will say to the contrary.

There is always someone who opposes everything that is done and if you mention the fact that you intend going into the pure-bred poultry business they straightway oppose you, and try and talk you out of it. And there are a lot of such fellows that when shown by actual results, will shake their heads, and say, "Well, it won't last, it was just an accident."

HOW TO MAKE A SURE SUCCESS OF THE PURE-BRED POULTRY BUSINESS.

The same general elements that are essential in all business to make a success of it, are necessary in the poultry business. Some of the writers have different names for it, but boiled down it is just pure "common sense." There is a certain force in every one, if exercised in the proper direction, will develop good judgment, which is doing the right thing at the right time.

First, select good foundation, stock or eggs, or both.

Second, don't try too many breeds at the start, but work into the business gradually.

Third, start according to your means. If short of funds, buy more eggs than stock and erect buildings and yards according to your pocketbook. Success does not mean expensive buildings and yards.

Fourth, study and learn all you can from other breeders, and those in the chicken business. If you stumble into anything you don't know, write me. I have had considerable experience and may help you.

Fifth, don't let any little back-set discourage you. Show your grit. You can do what thousands of others are doing, proving the poultry business is the best for the money involved, and are succeeding in poultry.

Sixth, know how to advertise and get a demand for your fowls, eggs, etc. Learn how to select, prepare, and handle your best birds. How to judge, how to fit for sale. There are many of these things in this book that you can learn, and there are other books you can secure. See our list of the best works on poultry. I have a book on advertising that is absolutely necessary to beginners. Get one. Above all, show your common sense and good judgment in all things pertaining to poultry and you cannot help but succeed, as thousands of others are doing.

COMMERCIAL SIDE OF PURE BRED POULTRY.

It is clearly proven that Pure Bred Poultry pays in a commercial way much better than common stock. Pure bred poultry requires less feed, lays more eggs and makes larger and heavier birds and commands a better price and has more ready sale.

The cost of getting a start is only a little more; that will be more than many times offset in the additional profits the first year.

And then the satisfaction of raising pure bred poultry is so much greater than common stock, which counts for much, as every one will take pride in handling pure bred stock to such an extent that it is bound to receive better treatment, consequently will do better and make more satisfactory gains and bring better profits.

I think it is advisable for a great many to combine pure bred poultry with the commercial side of the poultry business. You know full well that the stock and eggs you cannot sell for breeding purposes, you can put on the market at a good

profit. I do not think there is such a thing as "fail" in the chicken business, if you start right and use just good "common sense."

I am going to make a statement that cannot be successfully denied, being fully demonstrated by actual experience. It is this: poultry products from pure bred flocks command a good price nearly everywhere, and at a better price than common stock. The market price for all poultry products is steadily advancing. Poultry and eggs are one of the most staple articles in the world today. You need never fear an over supply or congested market. You can't go wrong to improve your stock by introducing new and better blood, either through fowls or eggs, or if you are not a poultry raiser now, getting a start of really first-class stock or eggs, such as I can furnish you, and going in for all there is in it. You can sell your best birds and eggs for breeding purposes, at good prices, and what is left, for market purposes. I am sure you will never regret it.

RAISE THE STANDARD OF YOUR EGG PRODUCTION.

The latest farm census gives the average production of the hens in the United States at 60 eggs per year. The average for Iowa is 65 eggs per year, which is the highest of any state.

The egg product sold for $15,000,000.00 last year in Iowa; suppose that the average was 195 or just three times as much as at the present, look at this vast sum that could be made in the State of Iowa alone.

Thirty million dollars more profit, and Iowa stands at the head of the list in the highest average of eggs.

Now this average is absurd among an intelligent people, it is just as easy to keep a hen that will lay from 200 to 250 eggs, as one that will only lay 60.

It is all in the breed and feed; of course you have to feed but you must have the hen—the natural egg machine to produce from 200 to 250 eggs per year.

Now we have developed such a strain of chickens by careful selection of breeding stock, choosing only those fowls that will commence to lay early and have proper functions of egg production strongly developed.

We save you lots of labor and cost in developing such a strain by offering them for sale, to start with or improve your present flock, both stock and eggs, for less money than you can produce them. I have made it a study and spent time and money in finding out how. I save you all this.

When you are starting a new breed or flock, start with mine; which will be several times more profitable than the average as raised.

It does not cost much more, but it is worth a great deal more.

THE BEST BREED.

This is much discussed and a question that is asked me every day.

There is no best breed. It is more or less a matter of choice and must be decided by the persons engaging in the business. But your conditions, situation and surroundings have much to do with the breeds best adapted for your needs. Read our description of the different breeds. It will help you to decide.

If you have a good market for eggs and you have the room, raise the egg producing varieties, such as the Leghorn, Minorcas, Hamburgs, Anconas, etc. If you wish broilers, roasters and meat, the heavy breeds are the best, especially if you are limited for room.

If you wish both, get some of the general utility fowls or get a breed of each of these varieties. Many do this. If you are undecided after reading our catalogue, write us in regard to the matter, telling full particulars as to location, room to devote to fowls and time you have to devote to the business and I assure you that I will take pleasure in helping you to decide.

My advice is not to try too many breeds at first. Get proficient in one or two breeds first.

I want you to succeed, even if I do lose a sale of several varieties, as I know a successful person is my best advertisement.

Write me your ideas as to what breeds you prefer or which you are interested in and I can help you.

POULTRY FOR MARKET OR EGG PRODUCTION. WHICH?

The demand exceeds the supply and production in all lines of poultry. We have to import eggs. There are not enough broilers or roasters to meet the demands as proven by the very high price they bring. There is never enough of the pure bred stock for sale, to meet the demand for breeding purposes in the spring, so there is no danger of overdoing the poultry business. The question of "which line of poultry shall I take up?" depends upon circumstances such as locality, adaptability, capital and inclination. Away from market and not much capital at command to start with, eggs would be advisable. If close to town and $200.00 to $1,000.00 to start with, equip for broilers or combine the production of eggs and meat, the latter being followed by most people and one can hardly fail to make it go.

VITALITY IN POULTRY.

I believe one of the most important things today for the poultrymen is to renew or build up their flocks by procuring stock of the highest point in vitality. Produce good strong fowls with good constitutions or eggs from birds of strong vitality. Not only should the feathers, form and markings be correct, but the fowls should have robust constitutions and strong vitality.

It will pay you to change your breeding stock quite often. Male birds at least every year and from a breeder that will furnish you stock or eggs of the highest quality as to vigor and hardiness.

A good way to do this is to procure eggs and in that way obtain some females to keep up the vigor of your flocks. It is a well known fact that all animals do

A bunch of Pure Bred Single Comb White Leghorns. Fine money makers they are.

well to change from one part of the country to another and this is especially true of poultry. They seem to do so much better than the fowls that were on the place that they come to.

But in changing it is highly important that you get stock from a reliable breeder that makes it a point to have stock of stronger vitality. I want to impress upon you the great importance of obtaining only such stock, and to you who are starting in the poultry business, much depends upon obtaining birds strong in vitality as well as correct size, shape, plumage and egg production.

I just wish to say that we are very particular to breed for vigor, hardiness and strong vitality, so we have birds of exceptional strong constitution. Range raised birds, fresh air houses, kept well cleaned, good fresh wholesome food and strict attention to all details have produced a race of fowls that cannot be excelled for vitality and will strengthen your flock or give you a start that will be both a pleasure and profit to own.

RAISING BROILERS FOR MARKET.

This is one of the most practical and profitable branches of the poultry business and can be carried on successfully in a limited space and without keeping a flock of fowls to produce the eggs as they may be produced from people who make a specialty of raising pure bred eggs for broiler hatching. By the use of an incubator and brooder with a good brooder house and a small house and yard in which the young chicks can get sunshine and exercise, one may raise large numbers of young chicks and dispose of them at fancy prices when from eight to twelve weeks of age. There is a great demand for this class of poultry, especially in the cities, and at very remunerative prices.

BUILDINGS AND NECESSARY EQUIPMENT.

You must provide good shelter—not necessarily costly or luxurious, but good warm houses of some kind. They can be made of straw, slough hay or sod, and many a thoroughbred chicken has been reared in quarters made of this material and brought their owners good profit. But it is not necessary to use cheap building material or temporary arrangements long, as poultry will soon pay for good substantial houses. We think the most economical kind of a house to build is one that is similar to several we have built on the Golden Rule Poultry Farm. We have examined lots of plans and have seen lots of different styles of houses and have also studied the matter quite a bit. And, from a practical standpoint.

You can have them any length you wish, making them 16 feet wide, which we have found to be the best. We show an illustration of one of these houses on this page. These are the dimensions and kind of lumber to use: Height of front, 4 ft.; back, 6 ft.; highest point in center, 10 ft.; from highest point to where the roof joins the building, 3 ft.; studding and rafters every 2 ft. apart; sill, 2x6, laid flat on foundation. Spike studding to sill. Make window size to suit yourselves. Often they may be procured cheaply from some one taking out of their dwelling and putting in larger ones or replacing the old fashioned small lights with the modern larger ones. Yellow pine is cheaper for dimensions lumber; fir or white pine for siding.

Build them facing the south, if possible. Put a brick foundation to keep out the rats or you can make a cement foundation by digging very narrow trenches where the foundation is to be, filling up with cement, the dirt making natural boxes and then box to a level on which to place the sills. Put in plenty of windows. Nothing takes the place of sunshine. It is absolutely necessary for the health and prosperity of your flock. Put cheap curtains on the windows at night to shut out draughts. The dropping platform is hinged to the north wall, three feet above the floor. It may be lowered to clean it. The perches or roosts are made like saw horses and are easily removed. Then you have a clear scratching place for the fowls in the day time. You can make the houses fancy, if you wish to put on a cupola, use fancy doors and windows. But just the way we show it, is the most practical, reasonable and best.

POULTRY LICE AND MITES.

How we get rid of them on Berry's Golden Rule Poultry Farm.

Lice and mites are the greatest enemies in many localities and are often the direct cause of most of the ills to which fowls are heir.

If your fowls get anything wrong with them, the first thing you do look and see if they do not have lice and mites.

A great many firms advertise a machine for ridding poultry of lice. We herewith give you directions, with the cut below, of a simple arrangement easily made and operated that is very effective, and we have one on Berry's Golden Rule Poultry Farm which we have used with great success.

We use just a common sugar barrel, costing ten cents, and the illustration shows how it is done. Can use any kind of good barrel. Place eight to twelve fowls in it at a time. Sprinkle a little Berry's Sure Insect Powder, say a heaping tablespoonful, close the door and turn the barrel slowly until it has revolved twelve or fifteen times. Little chicks can be treated in the same way, placing 25 to 40 in the barrel, according to size. When they first come out they may act a little dumpish, but soon recover, and no ill effects result from their ruffling up.

Our Sure Insect Powder advertised and priced on another page is much superior to any other for this purpose. You must kill the lice and mites that remain in the poultry house, coops, brooders and nests, and must keep them down by painting the runs, roosts, nests and walls with Berry's Sure Lice Killer. Do this three or four times a year and you will not be bothered with these vermin. Full directions for using on each can. This liquid is fully guaranteed by us and if it does not fully and effectually do the work get your money back.

THINGS WE DO ABOUT OUR YARDS AND WHAT YOU SHOULD DO.

Keep the houses well whitewashed inside.

Make the roosts not over 8 to 12 inches high above the dropping boards.

Give plenty of ventilation to your house. **But no draught over the fowls.**

Use Berry's Lice Killer freely on roosts; it keeps lice out of business in their strongholds.

Keep dust bath without fail, in this way fowls rid themselves of lice. Ashes mixed with Berry's Insect Powder kept in low boxes makes a splendid bath.

Don't fail to give male birds an extra feed; they need it during breeding season.

A little coal oil put on drinking water is good for colds and roup. Bordeaux is also good given in this way.

Have nests on the ground where water does not reach them. Eggs hatch well there. Put something over them to protect them from sun and storm.

Break up sitting hens as soon as noticed staying on the nest. Have an extra pen with a good cockerel you are keeping over for cock bird; throw the hens in this pen and in a few days they can be thrown back into the breeding pen.

SOME DISEASES AMONG POULTRY
ROUP IN THE VARIOUS STAGES.

One of the most dreaded diseases among poultry is that of roup, which usually begins with a cold. Fowls are subject to colds as well as humanity, and should have the same attention we would give ourselves, for we all know that should we neglect to apply a remedy when we take cold, the result might prove quite serious. The same will be the case should your fowls take cold, which may be brought about in various ways, viz.: Roosting in damp quarters, draughts of air passing over their sleeping apartments, sleeping in coops on the ground where they are packed so close together that some smother during the night, and those not suffocated are overheated, and when they get out in the cold air in the morning, a severe cold is the result, and if a remedy is not applied and cause removed, roup will invariably follow, which of all poultry diseases is most obstinate, sickening and hardest to cope with. If necessary precautions are not taken in the start to arrest the disease, it will run through the entire flock and leave nothing but death and destruction in its path.

Symptoms—Fowls begin coughing, sneezing and often their breathing is heavily accompanied by a wheezing sound. Their eyes become inflamed, heads swell and have a watery discharge from the nostrils, which sometimes leaves an offensive odor. They also drink almost continually if they have access to water, which is indicative of fever.

As the disease advances the head becomes swollen on one or both sides, frequently obstructing the eyesight. Often when fowls are affected with this disease they have a good appetite and will eat to the last, unless they are internally affected, in which case they are stupid, and a discoloration of their excrement may be noticed, which is much the same as that of fowls affected with cholera.

Cure for Roup.—When fowls are in the advanced stages of this disease the hatchet is the best remedy, as it is almost impossible to effect a cure. In the early stages they may be cured by taking a small spring bottom oil can and injecting a little kerosene oil in their nostrils and roof of their mouth. Also give them Berry's Roup Cure as per the directions on each bottle and is a sure cure or no pay. Also put a half teaspoonful of aconite in each quart of their drinking water. See that they are provided with good, dry, comfortable quarters, with an abundance of sunshine in their room, which should be well littered and changed frequently. Their drinking vessels, and vessels in which are fed their morning mash, should be cleaned with boiling water. This is absolutely necessary to accomplish a speedy cure. Be sure to remove all sick fowls from the ones not affected, to prevent the spread of the disease.

DIARRHOEA OR CHOLERA.

There are various causes for this disease—a change of food or a feed of sour meal may produce it. Indigestion caused by want of suitable food; liver complaint brought on by lack of exercise; improper feeding; inflammation of the intestines set up by disease germs; by the presence of worms; or, in fact, any one of a dozen causes may operate together to cause relaxation. Cold, and a spell of wet, cheerless weather will play havoc with fowls, if not well looked after.

The first thing in the way of treatment is to place the fowls where they will be perfectly dry and reasonably warm. Medically give the patient a dose of Berry's Cholera Cure. Full directions with each bottle. It is better to reduce the food supply, giving only a little oat meal boiled in milk, and frequently small doses of oil. In future see that the food is of good quality and keep the fowls in comfortable quarters. At present it is advisable to omit green food and to feed the birds almost solely on wheat.

WEAK LEGS.

Weak legs is a disease that attacks young birds, cockerels more frequently than pullets. The fowl is more or less incapable of holding itself up, and frequently sinks to the ground, and in severe cases is unable to stand. The health is otherwise good; so is the appetite, the bird not having suffered from want of exercise.

This disease, which sometimes attacks the finest birds in the flock, is caused by a too rapid increase of weight, out of all proportion to the development of muscle. It usually attacks the heaviest fowls, but is seldom found among old birds. It is common among the heaviest varieties. Constitutional weakness will also produce it, without rapid growth. Color drinking water slightly with tincture of iron. A good supply of nourishing food must be supplied, and it should be a kind calculated to produce flesh, not fat. Barley, millet, ground raw bones or chopped lean meat are beneficial. Plenty of fresh green food should be provided.

SCALEY LEGS.

Scaley legs is a disease in fowls caused by a small parasite, which can only be seen by the aid of a microscope; this parasite works under the scales or skin of the legs and puts them in very bad condition. The real cause for the parasite and scaley legs is letting the birds roost in unsanitary houses, such that are rarely cleaned. Keep your hen house as clean as your kitchen and your birds will never have scaley legs. A good cure is to wash the limbs in a strong solution of soap suds, then rub with coal oil and lard mixed, give the latter application about once a day for three or four days, then rub off all the scales that will come off easily and give the feet a good vaseline bath and they will be as smooth and clean as they should be.

CHICKEN CACKLES.

Keep everything scrupulously clean under roosts and in scratching pens. The litter in the scratching sheds becomes very filthy and too dusty or damp if left in too long. Clean litter in which to throw the feed is appreciated by poultry.

Never feed small chicks the first two days. Feed very lightly with cracked or rolled oats and wheat screenings, cutting green stuff or cabbage very fine every day or two. Never feed cracked corn or corn meal to them until two or three weeks old and you will not have much bowel trouble among your chicks.

When your chickens have rattling in the throat give them lard with some turpentine inwardly three or four times a day. Also rub well along bill under throat.

Know Your Birds

LEARN THE POINTS.

(A)—Comb; (B)—Face; (C)—Wattles; (D)—Ear Lobes; (E)—Hackle; (F)—Breast; (G)—Back; (H)—Saddle; (I)—Saddle Feathers; (J)—Sickles; (K)—Tail Coverts; (L)—Main Tail Feathers; (M)—Wing Bow; (N)—Wing Bar; (O)—Wing Bay; (P)—Wing Butts; (Q)—Breast Bone; (R)—Thighs; (S)—Hocks; (T)—Shanks; (U)—Spurs; (V)—Toes.

HATCHING TABLE.

Chickens	21 days
Ducks	28 "
Turkeys	28 "
Geese	30 "
Pheasants	25 "
Guinea Hens	25 "
Partridges	24 "

EXPLANATION OF TERMS.

Clutch—A setting of eggs or brood of chickens.
Cockerel—A young male less than one year.
Crest—A tuft of feathers on the head.
Cushion—The feathers which surround the tail in Asiatic breeds.
Deaf Ears—The ear lobes.
Dubbing—The removing of the comb and wattles.
Face—The bare part around the eyes.
Flights—The long quill feathers of the wing.
Fluff—The soft feathering below the tail.
Furnished—A bird is said to be furnished when it is fully developed in plumage and body.
Gills—The Wattles.
Hackles—The long narrow feathers on the neck.
Hen Feathered—A male bird is so described when he has the plumage of the female.
Hock—The second joint from the ground, intermediate between the foot and thigh.
Keel—The breast bone.
Mossy—Clouded markings.
Pea Comb—A triple comb.
Penciling—The narrow markings round or on a feather.
Primaries—The small back feathers on the wing which are concealed when it is closed.
Pullet—A young female bird.
Saddle—The short feathers on the back next the tail.
Secondaries—The hard feathers in the wing which show when it is closed.
Shafts—The quill of a feather.
Shank—The part of the leg between the foot and the hock.
Sickles—The long curved feathers in the male bird's tail.
Spangling—The dark spots on the feathers of certain breeds.
Squirrel Tail—A tail that is carried too much over the back.
Tail Coverts—The short feathers at the sides of the tail.
Vulture Hocks—Hard feathers attached to the hocks.
Wing Bar—The dark lines across the wings of certain breeds.
Wing Bow—The top part of the wing.
Wing Butts—The end of the wing.

A Flock of Our Golden Rule Poultry Farm Leghorns.

Golden Rule Poultry Farm

We show a bird's-eye view of the mammoth poultry breeding farm inside the first cover of this book. It is a plant that we are proud to show you.

This plant is conducted on the farm of Mr. and Mrs. A. A. Berry, Clarinda, Iowa. The farm consists of over one hundred acres and each variety is given plenty of range. On this farm all the business connected with our large poultry plant is conducted, eggs packed, fowls cooped, etc.

For a number of years past we have been breeding poultry, building chicken houses, coops, colony houses, laying out large and commodious yards, operating and testing brooders of different kinds, manufacturing incubators and brooders as well as poultry supplies and putting lots of hard work, study and money into the business and laying the foundation of one of the most complete pure bred poultry plants in the world. We expect to still increase and develop the plant. We have made it pay as we went along, buying the improved and very best well bred stock we could secure and made a good profit besides. We can start you and make it pay you.

In presenting this book to you, we have aimed to make it practical, truthful and a guide as to how you can make the poultry pay you; how you can turn dollars out of cents by simply offering you the choicest lots of pure bred birds, incubators, brooders and supplies that have ever been offered, from our Golden Rule Poultry Farm.

The Golden Rule Farm is all its name indicates and in all its dealings treats every one on the Golden Rule plan, which is: "Do unto others as you would others would do unto you." This rule enables us to put out strong, iron clad guarantees which, in substance, are as follows:

We fully guarantee everything we offer to be as represented and advertised.

There are no restrictions to this. How much fairer could anyone be? Ask yourself if you could do any better with those who buy of you or those you are dealing with. Our farm is not only Golden Rule in name, but Golden Rule in deed.

We are proud of our poultry plant. It is extensive, well arranged and up-to-date. We raise such poultry, make such incubators and manufacture such supplies as you may rely upon and which will make you money. In making our prices we have kept in mind the principle of the Golden Rule and made "live and let live" prices. We just simply cannot be excelled in the line of quality.

OUR PURE BRED POULTRY.

Quality has ever been our motto in building up a flock of pure bred poultry; quality first, last and at all times has been our aim in manufacturing incubators, brooders and supplies.

Fully realizing that much depends upon the start and foundation of a flock of poultry, we have spared neither time, expense nor energy in obtaining the best of high scoring birds that could be obtained. Not only high scoring, but big as to weight and great as to laying quality. Start right and you will always be ahead of the fellow who makes a poor beginning. You must have a good start of either eggs or birds to make a good beginning, and then keep on breeding from the best specimens raised or that may be procured, in an intelligent manner. We have bred poultry for the past twenty-four years and have always taken a deep interest and made a study of all lines of this business.

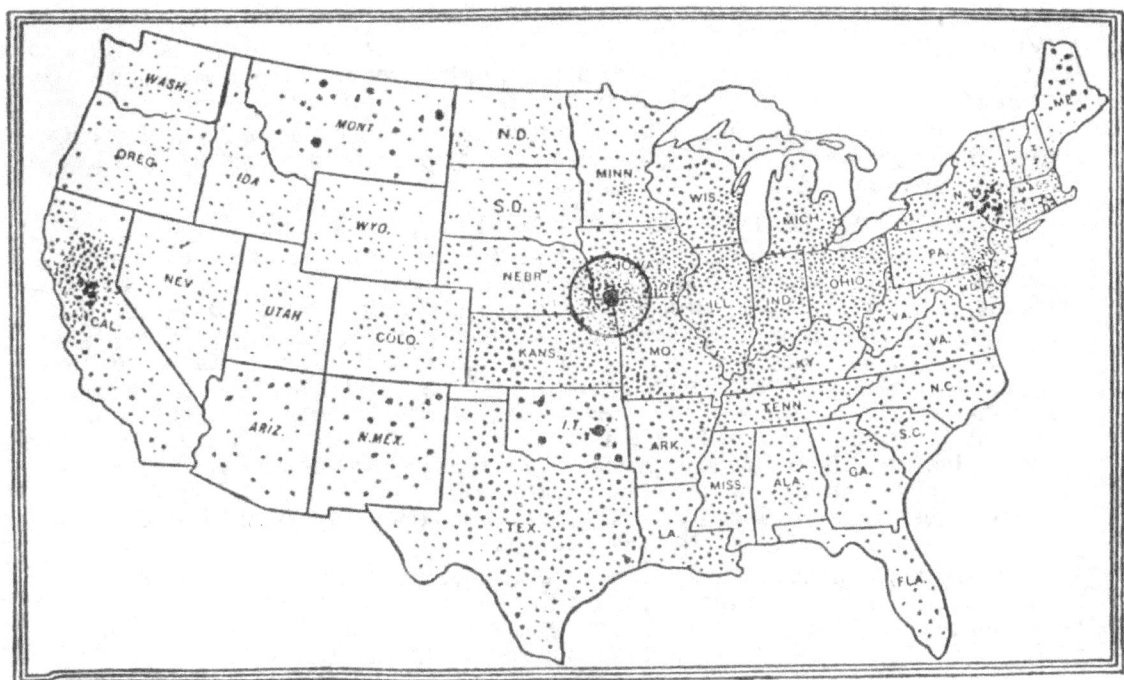

OUR LOCATION.

Southwestern Iowa is acknowledged as the greatest poultry producing locality in the world. There is more poultry of all kinds raised here than any other spot on the globe. There is more pure bred poultry raised than any other place on earth. The fowls do remarkably well and develop into magnificent specimens, strong, vigorous and with beautiful plumage. The climate is well adapted to develop strong, healthy, hardy birds, and feed is so plentiful that great numbers are always found on every farm.

The above illustration shows you where the poultry of the United States is raised, each dot represents one thousands dollars' worth of poultry and is very nearly accurate, although the records could not be obtained very accurately in some of the Western states and does not show as many as it should. But the map was made just as the last farm census shows and absolutely correct as the returns.

Note the little circle at the corner of the states of Iowa, Nebraska, Kansas and Missouri, and see how thick the dots are. Well, we live in the center of that circle so you see we are right in the midst of the poultry industry of America. We are headquarters for poultry and can furnish you with the best that grow and can afford to sell them for less money than others that are not so well situated. Please remember this when you are deciding as to where you will place your order.

ORDERS FOR STOCK AND EGGS.

In sending orders some have written us that if we could not furnish the variety called for asked us to send another variety, the party naming the second and third choice. Some stating that if we could not furnish certain variety to send the kind that would be best adapted to them, they telling us their situation and condition.

This was most generally in regard to eggs, specifying that they wanted them shipped at once.

And we think that is a very good idea for a great many who are starting in the poultry business as if they should tell us how they are situated as to room and buildings, time they could devote to the business, locality and demand for poultry products we could choose for them to their advantage.

So if you are not sure or particular as to the variety, please name a second choice or ask us to choose something suitable for you. Now please remember that this is for your convenience and not ours. If you only have the one choice we will take pleasure in filling your order.

ORDER EARLY.

We advise you to place your order as early as possible. While we have a large supply of fine birds now, our supply might become exhausted before the season is over. You can buy cheaper now than you can later.

On the following pages you will find a description with photograph of the varieties of poultry we offer. We hope you will favor us with an order. We know we can please you and are anxious to do so.

NEW BLOOD INFUSED IN OUR FLOCKS EVERY YEAR.

I am often asked the question as to whether I can supply cockerels for mating purposes for my customers, when they have purchased a start from me the previous year, or purchased breeders of any description. I most assuredly can, I refer to my records and find out just what strain was used in yielding your previous order, so I would chose such birds that would be best adapted to produce the very best results. I have a number of pens and flocks of each breed, it taking more flocks of some of the breeds than others, thereby changing my matings and infusing new blood, scientifically mating for the high standard that Berry's Line Bred Birds have attained. By infusing new blood every year I am eabled to supply all my old customers with cockerels that are adapted to their needs for best results. I can also furnish pairs, trios and pens of non related stock and properly mated up for best results.

My success depends altogether upon your success in this line of work and I assure you my dear friend that I use all the means at my command and my very best endeavor to provide you with birds and eggs that represent the best and of such quality and value that will make success a sure thing.

A WORD ABOUT PRICES.

Did you ever see prices of all poultry products so high? I don't think so, I know I never did. Isn't it just splendid the profitable prices that are paid for chickens and eggs on the market at present and has been the last year. I mean for those that have them to sell, but will admit it is pretty hard for the people who have to buy, but then there is nothing to hinder or prevent any one from going into the chicken business and getting a whack at those high prices.

Do you know that the market price has doubled in the last five years? Well it has most everywhere, but I have kept the price in my annual book, "Profitable Poultry," about the same while I have improved the stock and quality at a big expense, and then feed is so much higher now and it costs more to advertise, so I have advanced the price a little on some items this year, not very much and not on a great many. I have not advanced the price in proportion to the cost of producing them.

You will find the price lower than others, if not, write me, I will meet all prices of equal value, in fact I doubt if you can produce quality at anything near the price I am offering my superior and high class stock. Don't let the price worry you. You have no trouble in getting the price when you sell. Look more to value, no trouble about the prices as they are going higher all the time.

QUALITY FIRST CONSIDERATION.

I have always considered quality above everything else, but I am situated so that I can give the highest quality at lower prices than any other that I know of Don't let a little matter of price stand between you and getting some real fine, well bred stock or eggs, it will pay in the long run. At the Boston show not long since a Plymouth Rock hen sold for **seven hundred and fifty dollars.** Down in Missouri there is a man that got **seven thousand five hundred dollars** for a pen of White Orpingtons, he also made a profit of over eighteen thousand dollars from his chicken farm last year. At the Indianapolis show a **White Plymouth Rock cock sold for eight hundred dollars** and eight other birds from the same yard for **nine hundred and fifty dollars** making a sum total of **one thousand seven hundred fifty dollars** for nine birds all from one man's yard. A young man saw a nice young cockerel in a farmer's barn yard and bought him for five dollars, he took this bird to the poultry show and in less than one hour sold him for **fifty dollars.** There are a great many birds that sell from five dollars to a hundred dollars every year, any man who breeds good birds will find a profitable market for them at home. I have a customer, a young widow woman, who bought a setting of pure bred eggs, 7 years ago, and began to breed fine chickens; she did not have much notion of making more money from them than from the scrub fowls she had always kept. She had a nice house and a well kept place and wanted her chickens to look alike and be a uniform size and color, she also wished to increase her profits to help in her maintenance and in raising her family, it wasn't long before her neighbors discovered that she was getting more eggs from her pure-bred hens than they were getting from her scrubs, they soon discovered that quality made the difference and begun to buy eggs for hatching and surplus cockerels and pullets from our friend. In one year she made $899.25 from 185 hens, this made an average of $4.85 each, she did not advertise, her chickens and eggs did not cost her one cent more than they would if she had kept scrubs, but the quality of her fowls sold them and caused such a demand that she got 75 cents to one dollar a doz. for all the eggs she could supply to her neighbors for hatching, she gets $2.00 each for cockerels, $1.50 each for hens, you could not give her a scrub hen now as she has been making big round dollars for several years, for every cent she invested in the first place.

I have made such a close study of breeding the best fowls that I know just what they will do. I have been selecting the best layers for years, egg production is the first thing, fine feathers are all right as far as they go, but they do not go far enough for me, I have bred poultry long enough to know points of good well bred birds when I see them, I study my fowls until I know what each one is doing and the loafer is disposed of the minute she is discovered.

I CAN UNDERSELL ANY SO-CALLED SPECIALIST.

How is that you ask, because where the specialist breeds a dozen I breed a hundred. I do business on a big scale and where the specialist is asking $5.00, $10.00, $20.00 or even $50.00 for a bird, or eggs in like proportion, I can give you birds equal or better in quality from $1.75 to $5.00, so do not stand back in purchasing or sending me your orders if you think that the quality is not good enough because I do not price my stock so high as many of the specialists. I can furnish quality and sell at listed prices and make profit enough to suit my modest ideas, and the "Live and let live plan I follow."

IT WILL PAY YOU TO ORDER FROM THE GOLDEN RULE POULTRY FARM.

You can get the best from us at a reasonable figure and at a price that cannot be duplicated.

We have the fowls and can supply you with them or eggs from them.

I will make this assertion and am open to the closest investigation, that we have more pure bred birds on our farm and have more under our supervision of the various specialty flocks on different farms about Clarinda than any other specialty poultry breeder and incubator manufacturer in the United States. Ours is not imagination or exaggeration; but a reality.

We can fill your order with better fowls or eggs for less money than any other firm, as we have a larger and better supply to draw from.

MODERN VERSION OF THE GOLDEN RULE.
A Square Deal.

In the fifteen years that I have been selling pure-bred poultry, eggs, incubators, etc., I have never taken a person's money and sent out an unfair equivalent, nor have I taken advantage of me having another person's money and could send any old thing.

Aside from the moral phase of the matter (and I will just say that I am a church member and teach a Sabbath school class of young ladies)—just as a business proposition I ask you if I have not "gouged" folks when I was in debt and struggling hard with work and worry would I be likely to do so now when I have so prosperous a business and on "Easy Street," as the saying is?

I think not. What do you think, Sister, Brother? Say, let's get acquainted and do business together. This splendid verse from that grand old poem of Waterman's expresses my sentiments. What do you think of it?

"If I knew you and you knew me,—
If both of us could clearly see,
And, with an inner sight, divine
The meaning of your heart and mine—
"I'm sure that we would differ less
And clasp our hands in friendliness;
Our thoughts would pleasantly agree,
If I knew you and you knew me."

SHIPPING FACILITIES.

Our shipping facilities are perfect, we can ship to any part of the world. We have received orders from not only customers in the United States, but from Canada, Mexico, the Hawaiian Islands, Japan, West Indies, South America, etc.

There are six different branches of railroads running into Clarinda, and we can ship out on any of these in any direction, so that the birds or whatever goods you happen to order will not have to travel hundreds of miles out of the way before they reach you. But we are able to send direct, just whatever direction you happen to live.

Two of the leading express companies in the U. S. have their offices here, and these two make direct connection with all others. A year or so ago each company made an extra charge for transferring to another company, so that if your goods would happen to go over two companies you would have to pay a little more than if one company would handle it all the way through. But that ruling has been entirely set aside, and a new ruling has taken effect, and it does not make any difference how many companies handle the goods, the charges are fixed according to the distance, and there is no extra charge for transferring from one company to another. We thereby can arrange to get for you the very closest express rates that it is possible for anyone to get, and since we ship so very much, the express companies cut the rates down just as close as they possibly can because they know that if they mistreat our customers, we will not give them the business that is due them.

We show here a couple of photographs of our express company wagons at our shipping rooms, loading up shipments for our customers. In the busy season these wagons come out two and three times a day, and take from 20 to 40 coops in a load each time. This simply shows that we are well known everywhere, and it does not take people very long to find out where they can get the very best at the very lowest prices.

I wish that when you send me your order you would give me the name of your express company or express companies that have offices at the point where you wish goods to be shipped, as it will save us a whole lot of work in looking up the matter. This must be done so that we can properly arrange our billing.

We wish to call your attention to the express rates that were given us by our express agents, found on page 127. You will also notice on the preceding page what we have to say in regard to the Parcels Post shipments. Give this careful consideration also. Shipments are fully guaranteed no matter how you order them shipped. We assume the entire responsibility. There is one assurance that we can

give you, and that is that you will never be disappointed with anything that we would send you and you will never have occasion to regret having placed your order with us.

This is a picture of some of our big shipping houses, the larger one with full basement is 50 by 30 feet and will hold 900 chickens. The other 30 by 40 feet will hold 600 chickens. The smaller is a colony house for 50. Our boys are coming in with a basket full of eggs.

This picture shows a portion of the interior of the above shipping houses. Note the way we handle and prepare our birds before shipping. The coops are the best and most sanitary. We caught George in the picture, feeding the chickens.

OUR SUCCESS IN POULTRY BUSINESS.

We have always been very successful in the poultry business; so much so that this elegant and useful book is the outcome of our efforts in that line. We have pleased our thousands of customers so well in filling their orders that the business has developed into a great industry. We have put lots of capital in stock, fixtures and appliances of our plant, lots of energy and study in the breeding of thoroughbred stock and are well prepared to fill your orders from the varieties we advertise in this book. You can always get the best from us for a reasonable figure and at prices that cannot be duplicated.

The Great Secret of Success

Advertising the Poultry Business

HOW TO DO IT PROFITABLY.

I have written a little manual on advertising the Poultry Business, that if followed cannot help but dispose of your stock and eggs at a good profit.

Do you know that it is more of an art to sell than produce? It takes more brains—more know how to advertise than to raise chickens; but with this poultry advertising manual you cannot fail to sell all that you can raise at a good figure. It is plain and practical and every one can understand and follow its instructions.

It is the experience of one that has made good and one that knows how. It contains information and secrets that are worth hundreds of dollars to every one going into the poultry business. It has cost me thousands of dollars in actual money spent in newspaper advertising.

Now, my customers' success is my success. I will succeed a great deal better if you do well in the poultry business. Some cannot help but succeed, while some are not as well situated or do not grasp all the little details that are necessary in advertising.

In order that you may succeed with poultry, I am going to send you a copy of the advertising manual with each order amounting to $5.00 or more. This manual is not for sale, it is too valuable to let every one have a copy even if I charged a good price. It is only for my customers that I am putting this treatise in their hands to make success a certainty and much more successful than they would be without the manual.

I do not have space to devote to the subject here, but have gone to the expense of treating the subject in book form.

To all those in the Pure Bred Poultry business and to those that are going into that profitable and interesting business this manual is of great assistance.

And it don't cost you anything if you send an order, as it is for the exclusive use of my customers. It tells how to write advertisements that bring results, it tells what to say in advertisements, it tells what papers are best to use, gives the price of advertising space, the circulation of papers, it explains the classified advertising, it tells how much to spend in advertising—how to write and get up a circular or booklet that will get business, and gives a lot of valuable information that is absolutely necessary.

This book is strictly for the benefit of my customers, and is given with each $5.00 order upon request. The price of the book is $1.00, but is worth a great many dollars, as it will both make and save a great many dollars for everyone who has pure-bred poultry or eggs to sell.

However, if some of our customers do not wish to make their order for $5.00 and yet want this great Advertising Manual, I will mail it upon receipt of $1.00. If after receiving the book, you do not think it is worth $1.00, you can return same, and I will send your money back. Fair, isn't it?

BERRY'S GOLDEN RULE POULTRY FARM, CLARINDA, IOWA

The Following Pages Describe, Illustrate and Give Prices of Our Fowls

BARRED PLYMOUTH ROCKS.

The above cut represents one of our Barred Plymouth Rock cockerels as much as a picture can represent anything. Isn't he a fine fellow? Well, ours are all fine and will please you. We can furnish you fine cockerels in quantities at bargain prices.

It is hardly necessary to give a long description of this magnificent, all-purpose and popular breed of chickens, as they are probably better known than any other breed. The breed is as solid as the name indicates and they stand acknowledged as the best general fowl in the world today. They are very rapid in development and make excellent broilers at the age of 8 to 10 weeks. They are very good layers. They are an American bred bird and every good citizen should be proud of this grand distinctive breed for American climate and conditions.

In color they are the old dominique or cuckoo bars, white and black across the feathers. In the highest type of pure breds these bars are very pronounced and quite distinct, while in the mongrel or cross bred these barrings do not show distinctly, but run together, murky and brownish and the females are quite dark.

For a good many years the pullets were inclined to come pure black, while the cockerels were inclined to be light. This led to double the mating system when a pen would be mated to produce good cockerels and another pen mated to produce good pullets.

A Pair of Our Barred Plymouth Rocks.

There is no doubt but what the Barred Plymouth Rocks we offer, are superior to almost any others as this is the breed by which we obtained our reputation and have carefully bred this all-purpose fowl for the past twelve years. It is the breed that led us into the pure chicken business. Through this most popular of all varieties, we discovered the great possibilities of the poultry business. We do not desire to laud this variety to the hurt of the other varieties we are now raising and offering as they are all of the best and no one has better or can procure better. But, I wish to emphasize the fact that the Barred Rocks are our "first love" and they certainly fill the bill as an all-purpose fowl.

If you are breeding this wonderful breed and wish to secure some new blood that will improve your flock, do not hesitate to send to us for either eggs or cockerels, or if you wish to get a start you cannot do better than to send to us. If you have an eye to your pocketbook and want the best you will make no mistake in sending to us.

Shortly after commencing to breed this excellent fowl we decided that we preferred Plymouth Rocks of the single mating to the double mating and by selecting birds of the standard color and culling out the chicks that were off color, have perfected a strain that would produce good cockerels and pullets from the same mating. The great trouble in the color of a Barred Plymouth among the general class, is their tendency to become dark and murky and most flocks will degenerate that way unless care is taken to prevent.

We have always aimed to keep the barring perfect, distinct and light with genuine blue ringlet bars and if you want something that will improve your flock or wish to start with the Real Thing come to us. We have a large tract of land, with houses, pens and range devoted to our Barred Rocks and can furnish stock or eggs that will make you prize winners.

We give special attention to egg production in our Barred Plymouth Rocks and have not sacrificed feathers, form or shape for this great qualification.

PRICE.

We have three pens of No. 1 mated with as fine cockerels as there is, procured at a great cost. They are wonderfully fine blue ringlets, barred to the skin on every part of the body, splendid proud carriage which is typical Rock shape, short yellow legs, bay eye, yellow beak and skin and is as well nigh perfection as there is. The females were selected with great care and are fine specimens. If you want to raise some fine cockerels to head your pens these will do it.

Price: Eggs, $2.00 per 15; $3.50 per 30; $4.75 per 50; $8.00 per 100.

Pens 2 and 3 are perhaps better than most breeders first pens; they are composed of an equal number of pullets and hens with high scoring cockerels. They have fine even barring which runs to the skin, they are good sized, splendid birds.

Price. Eggs, $1.50 per 15; $2.75 per 30; $3.50 per 50; $6.50 per 100.

Pens 4 and 5. We have several pens of this grade while they are not quite so fancy and high scoring as the others they are better than the average Barred Rocks as found among the breeders.

They have nice even deep barring free from brass and defects and in all will please you.

If you are looking for incubator eggs or first-class high grade birds for farm or to start a flock get some of these.

Price: $1.25 per 15; $2.00 per 30; $2.60 per 50; $5.00 per 100; $9.00 per 200.

Price of Stock. Very choice ones, high scoring males, $3.00 to $5.00 each. Very fine ones, $2.50 to $4.00 each. Splendid ones, $1.75 to $2.25 each. In lots of two or more: Pairs, $5.00; trios, male and two females, $7.00; pen, male and four females, $11.00; pen, male and 10 females, $22.00.

SPECIAL MATINGS FOR EXHIBITION AND EXTRA FANCY.

We have a special mating pen of a number of varieties. They are the fancy birds picked out from my first and choicest pens; these are what I use to raise my breeders from and from these pens come the exhibition stock, that is, there will be more exhibition birds in proportion to the number hatched. These pens are the Vanguards, the leaders in the front ranks of the best and most improved birds in the way of size, shape and color and above all in egg production. I not only keep up with the times in procuring the best of all varieties we handle; but right in the front and on top in quality. These special pens are how I do it and enables me to have better birds and furnish better eggs from my pens, yards and flocks than others. We are Leaders. We are making the business a success. We are adding new buildings and new pens, enlarging and improving what we have and are nearing the goal of the greatest pure bred poultry breeding establishment on earth. Now I can spare only in setting lots of special matings from the following breeds:

- Partridge Rocks
- Barred Rocks
- White Rocks
- Buff Rocks
- Golden Wyandottes
- Silver L. Wyandottes
- Golden L. Wyandottes
- Columbian Wyandottes
- Partridge Wyandottes
- Buff Cochins
- Light Brahmas
- Black Langshans
- White Langshans
- Single and Rose C. R. I. Reds
- Cornish I. Games
- Black Minorcas
- Hamburgs
- Houdans
- Buff Orpingtons
- Buff Leghorns

Price of Eggs, $2.50 per 15; $4.00 per 30. Exhibition birds are a matter of correspondence.

Copyrighted 1905

WHITE PLYMOUTH ROCKS—Fishel and Heck Strains.

This is a well known breed and it is not necessary to give detailed description of them. They have come to the front very rapidly in the past six years. They are snow-white in plumage, have beautiful red combs and wattles.

These are very much like the Barred Rocks except in color and a great many regard them as being superior to that breed, claiming that they are better as an all around fowl or for broilers. Splendidly adapted as market fowls, are good layers, steady and reliable setters and careful mothers. In fact, there does not seem anything to hinder them from being great favorites as they surely are now.

They are easily fattened at any time of the year and their flesh is of firm quality and so deposited that it is largely laid on the choicest parts of the body.

They are the breed greatly recommended as soft roasters, a very profitable branch of the poultry business practiced in the East.

We have several breeding pens and the birds are all selected with very great care, headed by large snow white, strong, vigorous, deep bodied cockerels and we expect to hear from many prize winners in the show ring from these, as we have since we commenced to breed them. If you wish the best White Rocks that may be secured, send in your order for stock or eggs. Note the very reasonable prices.

Price: Eggs, $1.50 per 15; $2.75 per 30; $4.00 per 50; $7.00 per 100; $12.00 per 200.

Price of Stock: Single, $3.00; pairs, $5.00; trios, $7.00; one male and four females, $11.00; one male and 10 females, $22.00. Special price on cockerels in large numbers. We can guarantee you extra value for your money. Let us know your wants in this line.

SILVER LACED WYANDOTTES ARRIVE IN FINE CONDITION AND ARE NICE BIRDS.
Dear Madam:— Hale Center, Tex., 7-8-'13.
   The pen of Silver Laced Wyandottes arrived today in excellent condition and am well pleased with them. They are perfect beauties. Thanking you for such nice birds and prompt shipment, I remain. Yours truly, R. W. PATTERSON.

BUFF PLYMOUTH ROCKS.

Practically a new breed and like the Barred Rocks in every way except in color. It is a color that has always been a favorite with poultry breeders and as the Barred Rocks were a success and such a general favorite strain, this new strain made friends very fast right from the first and has become very popular.

In color the Buff Plymouth Rocks are a bright cinnamon in all parts of the body, the plumage of the cockerel being very bright, reflecting the light in the sun, while that of the hen is not so lustrous. They are the same size, style and shape as their Barred Rock cousins. The legs, beak and skin are of a rich orange yellow, while their combs, wattles and ear lobes are a bright red. This makes them very attractive. Our stock of Buff Rocks is A1 and score high.

They cannot fail to please you. Observe our low prices for this breed.

Price: Pens No. 1 and No. 2: Eggs, $1.75 per 15; $3.00 per 30; $4.00 per 50; $7.00 per 100.

Pens No. 4 and No. 5: Eggs, $1.50 per 15; $2.50 per 30; $3.75 per 50; $6.00 per 100; $10.00 per 200.

Price of Stock: Single birds, $3.00; pairs, $5.00; trios, $7.00; pen, 1 male and 4 females, $11.00; 1 male and 10 females, $22.00.

Fine cockerels in quantities at a bargain. Write for special prices.

For special mating see page 35.

A WORD ABOUT OUR DIFFERENT PENS.

They are all good and bred up for fancy points; but above all for the greatest egg production. The first pens are somewhat higher scoring and perfect in all the details, but the third and fourth are as good as the average breeder's first pens. We have them mated to produce best results of egg production and strong hatching qualities. I know that you will be pleased and hatch out many prize winners.

WE ARE THANKFUL FOR PROMPT SHIPPING.

Dear Friend:— Chicago, Ill., 6-4-'13.

I thank you very much for your prompt shipping of my order of eggs, which arrived in perfect condition. Not one was broken. We expect great success with these eggs.

Yours respectfully, CHAS. ROCKSTAD.

PARTRIDGE PLYMOUTH ROCKS.

This is one of the newest breeds and is certainly a splendid acquisition to the feathered tribe.

The Partridge Plymouth Rock combines the beautiful rich partridge color, with its wonderful contrast and even markings and trueness of type that is surprising with the shape and form of the Barred Plymouth Rocks that are so greatly admired. I always admired a partridge colored bird, they are to my eye truly beautiful and handsome and they are good. Plymouth Rocks have held first place in the affection of the American people for years past; they still hold the first place in numbers of all the different breeds, but now there is only the color in the difference between the Barred and the Partridge Plymouth Rocks and I am sure you will just love the partridge color as they are truly beautiful and they are a brand new breed, one of the latest and so firmly established that they breed truer than many of the older breeds. Being such a new breed and as the type is so strongly established, they are wonderfully strong and vigorous. They have the true Rock shape so popular everywhere, yellow legs and quick maturing are among their characteristics. Wonderful layers and especially good winter layers, they are the ideal all-purpose fowl combining the meat and egg production.

There has been no fowl that has become so popular in so short a time, because they are a perfect combination of yellow legs, plump bodies, splendid for fries or broilers, rapid growth, great laying qualities and large size, together with great beauty.

Get in front of the procession with this splendid new breed.

Price: Eggs, $2.25 per 15; $4.00 per 30; $6.00 per 60; $10.00 per 100.

Price of Stock: Single birds, $3.00; pairs, $6.00; trios, $8.00; pen, 1 male and 4 females, $13.00.

WHITE WYANDOTTES.

Next to the Barred and White Rocks, probably the White Wyandottes are the most popular breed of chickens and many contend that they are the best all around general purpose fowl. There is no doubt that they are the most popular breed for broilers and are used on the great broiler plants of the country.

They are pure white in color, feather out quickly and cover with plumage at a very early age. They have very square, compact, plump bodies. They have a neat, rather small rose comb; wattles are small and ear lobes bright red. They have yellow legs and beak.

They excel as layers and especially as winter layers and respond readily to good treatment or do excellent as foragers.

For the past few years this breed has crowded all other varieties in numbers shown at the big shows and there is no question of their growing popularity.

We are proud of our fine flock of White Wyandottes of handsome and high scoring birds that we are sure will please the most particular customers. Our prices cannot be duplicated when the quality and high scores of the stock are considered.

Price: Pens 1 and 2: Eggs, $1.75 per 15; $3.00 per 30; $4.00 per 50; $7.00 per 100; $12.00 per 200. Pens 3 and 4: Eggs, $1.50 per 15; $2.50 per 30; $3.75 per 50; $6.00 per 100; $10.00 per 200.

Price of Stock: Single birds, $3.00; pairs, $5.00; trios, $7.00; pen, 1 male and 4 females, $11.00; pen, 1 male and 10 females, $22.00.

We can offer you some special bargains in cockerels in quantities if you will write your wants. Let us hear from you.

DUCK EGGS HATCHED FINE.

Dear Mrs. Berry:— Hamilton, Ill., 5-11-'13.

Am very well pleased with the fawn Indian Runner duck eggs I received of you. I hatched 10 ducks out of the setting of eggs.
Yours very truly,
JOHN HAMMON, Route No. 1.

BARRED ROCKS MARKED PERFECT.

Dear Madam:— Hot Springs, Ark., 9-17-'13.

The pair of Barred Rocks arrived in the best of condition and will say they are far better than I ever expected. The markings are perfect in every respect, and I surely appreciate the fact that you made the selection.

Trusting that you are doing a good business and thanking you for the good treatment which I received from you, I am. Yours truly, W. M. METZER.

P. S.—Anything that I may be able to do for you in the way of advertising, I will gladly do it.

SILVER LACED WYANDOTTES.

The oldest variety of the Wyandotte family. A very popular American breed and for a table fowl are unexcelled, having a very fine grained meat. They are among the best layers, careful setters and good mothers.

The cock is marked exactly like the dark Brahma cock on the back and wings, hackle and saddle, but the breast and body are black and white; the center of each feather being white, around this being a narrow lacing of intense black.

The hen has a silvery white hackle with a black stripe down the center of each feather. The breast and back are covered with feathers that are white in the center, laced or edged with a clear deep black border or lacing. The main tail feathers are a dull black on the upper side and white on the lower, the secondaries forming a distinct bar across the wings.

This breed has very square, compact bodies selling in the market for the highest prices, to those who want the very best. They have a neat, rather small rose comb. Wattles are small and ear lobes of bright red.

If you wish something fine and something of which to be proud and will win, get your order to me early and you will make no mistake.

Price: Pens 1 and 2: Eggs, $1.75 per 15; $3.25 per 30; $4.00 per 50; $7.00 per 100; $12.00 per 200. Pens 3 and 4: Eggs, $1.50 per 15; $2.50 per 30; $3.75 per 50; $6.00 per 100; $10.00 per 200.

Price of Stock: Single birds, $3.00; pairs, $5.00; trios, $7.00; pen, 1 male and 4 females, $11.00; pen, 1 male and 10 females, $23.00.

Have some extra fine cockerels to offer in quantities. Write for special prices.

THANKS US FOR OUR CONSIDERATION FOR CUSTOMER DURING HIS SICKNESS.
Dear Mrs. Berry:— Billings, Mo., 7-7-'13.
I received the Geese and Ducks, Saturday, July 5th, all in good shape and must say am more than pleased with them. I thank you very much for your kindness shown me while sick so much. I shall remember you when in need of anything in your line.
Very respectfully, GEO. HANNEBAUM.

BEST APPEARING EGGS EVER PURCHASED.
Dear Friend:— Gower, Mo., 6-16-'13.
I write you again in regard to the 100 Plymouth Rock eggs you sent me. They came in fine shape, only one broken, and a fine lot of eggs to look at, and as nice as I ever bought. And since I have been in the poultry business, I have bought eggs from parties far and near.
Yours respectfully, MRS. MARY WHITSON.

GOLDEN WYANDOTTES.

The Golden Wyandotte is marked exactly like the Silver variety, but the colors are different. Where the Silver Wyandotte is white the Golden is a deep mahogany red. This color makes the variety the handsomest one of the breed, the cocks especially being magnificent birds, their plumage having a metallic sheen that makes it appear as if it were polished. This variety breeds remarkably true to color. Birds of this breed are inclined to grow to a slightly heavier weight than those of the other varieties of Wyandottes.

This breed has many friends and is a good variety for any one to breed who wants beautiful fowls. There are very few breeds that will breed as true to color as the Golden Wyandotte. They make good mothers and good setters, but are not persistently broody; chicks are quite hardy, grow fast and mature early. My foundation stock was purchased direct from the best breeders in the East whose birds have won hundreds of prizes at the leading shows. In general utility Golden Wyandottes are much the same as the other varieties of Wyandottes, the only difference being in color. If you wish to start with the best stock in the country you will make no mistake if you order your eggs or stock from Berry's Golden Rule Poultry Farm.

Price: Pen 1: Eggs, $2.00 per 15; $3.75 per 30; $4.50 per 50; $7.50 per 100; $13.50 per 200. Pens 2 and 3: $1.75 per 15; $3.25 per 30; $4.00 per 50; $7.00 per 100; $12.00 per 200.

Price of Stock: Single birds, $3.00; pairs, $5.00; trios, $7.00; pen, 1 male and 4 females, $11.00; 1 male and 10 females, $23.00.

Can offer some fancy cockerels at bargain prices. Write for special prices in quantities.

A MISFORTUNE, BUT VERY GOOD SUCCESS.

Dear Mrs. Berry:—   Montezuma, Ohio, 6-13-'13.
I received the two settings of Partridge Wyandotte eggs in due time and set them. One of the hens got sick and spoiled one setting for me. The other hen hatched out 11 nice chicks. I feel sure that the other hen, had she had an equal chance, would have a lot of nice chicks to start with. I am pleased with them. Wishing you success, I remain.
Yours truly,   MRS. DAVID HEINLEN.

BIRDS ENJOY BUGGY RIDE.

Dear Mrs. Berry:—   West Pawlet, Vt., 8-4-'13.
I received the hens and cockerel in good shape. They are fine birds and show good care. They seem to enjoy the buggy ride as they sang and chatted on their way home. Thank you for your promptness, and you will hear from me again later. I am shipping the coop back today.   Yours truly,   MRS. MINNIE M. ROBERTS.

BUFF WYANDOTTES.

The above illustration represents the Wyandotte breed in the buff color; many prefer this color, and with it you get all the quality that is given for any of the other classes of Wyandottes.

In general utility, the size, shape, egg laying qualifications, etc., are practically the same in all the Wyandotte family, of which there are quite a number. You will notice that we list the Buff, Golden, White, Silver Laced, Partridge, and Columbian. Some prefer one, while others choose another. There are so many people who prefer the buff color to any other, and would like to have something in the Wyandotte class, and to lovers of such a fowl we highly recommend the Buff Wyandottes as above illustrated. It is a breed that will not by any means disappoint you. They are all that you can possibly expect of them. The Buff Wyandottes are good layers, and especially if you get a good laying strain introduced. We are able to favor you with just such a strain, and in fact no other strain should occupy your yard.

The Buff Wyandottes make good mothers and are good setters, but they are not persistently broody. The chicks are hardy and grow vigorous, maturing very early. Cock-birds will weigh from 7 pounds to 9 pounds, and hens from 6 pounds to 8 pounds.

The standard requirements of the color of this breed calls for a rich golden buff. We can furnish you with eggs of just such individuals, or can favor you with breeding stock in either males or females, so as to infuse new blood in your flocks, and thereby bring them up to what the standard requirements call for. Our pens are complete and we can furnish you with A No. 1 stock or eggs.

We know from the many letters that we get that the buff colored chicken is a very popular one, and also know this from the fact that the orders favor that color to a great extent. Here we have this beautiful color linked with rose comb variety that is so much desired among critical breeders. This breed to my mind is one of the leaders and I highly recommend them. We have a nice pen of them and can supply your wants.

Price: Eggs, $2.00 per 15; $3.50 per 30; $4.00 per 50; $7.00 per 100; $12.00 per 200.

Price of Stock: Single birds, $3.00; pairs, $5.00; trios, $7.00; pen, 1 male and 4 females, $11.00; 1 male and 10 females, $23.00.

PARTRIDGE WYANDOTTES.

The above cut is a good representation of a pair of our Partridge Wyandottes.

They are one of the coming variety of the Wyandotte family and are very popular; indeed, they are the choice of a great many practical poultrymen. The partridge plumage in any kind of a bird is handsome and they are noted for their beauty as well as usefulness.

These Wyandottes have a beautiful mahogany plumage with lacing the same as the Partridge Cochins.

In shape and size they are like all other Wyandottes, they are splendid layers, quick maturity and a very fine table fowl; they lay a good sized egg and will lay the entire year except when moulting, if proper care is given them.

They are hardy, easy raised, matured early and very popular for early broilers; the hens are good setters and make good mothers, although not persistently broody.

This breed is very attractive and possesses and combines all good points in placing them on the front ranks of a good all around fowl. We have a splendid lot of these beautiful and practical birds and can please you when you send in your order for fowls and eggs.

Price: Eggs, $2.25 per 15; $3.50 per 30; $5.00 per 50; $9.00 per 100; $14.00 per 200.

Price of Stock: Single birds, $3.00; pairs, $5.00; trios, $7.00; pen, 1 male and 4 females, $12.00. Pen, 1 male and 10 females, $24.00.

Special low prices in quantities of our cockerels to mate your flocks.

COLUMBIAN WYANDOTTES.

The Columbian Wyandottes are a very popular fowl indeed. It is a fowl that is not known by every one, and a fowl that you do not find in every barn yard. It is to my mind one of the most beautiful fowls.

You will notice by the above picture, that the markings of this bird are practically the same as those of the Light Brahma. The Light Brahma is the older of the two fowls, and being very beautiful, it was the desire of breeders to get a bird with the same markings with smooth shanks. The result was, the Columbian Wyandotte. This bird is not as large as the Light Brahma, by any means, and does not have quite the same shape. It has smooth, yellow shanks.

The Columbian Wyandottes are not unlike other Wyandottes as far as shape and other utility characteristics are concerned. The Wyandotte family, next to the Rock family, is one of the most widely known of any of the poultry breeds, and is looked upon as having peculiar characteristics of their own, that really excel some of the other well known breeds that are offered. The Columbian is by no means a sport of any of the other Wyandotte varieties, but it is a breed that has been built up by itself, and has been bred up from some of the most noted strains, and has proven to be one of the highest class of the Wyandotte breed. Our demand for this magnificent breed exceeds the supply every year. So do not lose sight of the fact that early orders will always bring you better birds and insure prompt attention.

Price of Eggs: $2.25 per 15; $4.00 per 30; $5.00 per 50; $8.00 per 100; $13.50 per 200.

Price of Stock: $3.00 per single bird; $5.00 per pair; $7.00 per trio; 1 male and 4 females, $11.00; 1 male and 10 females, $23.00.

RECEIVED EGGS IN SPLENDID SHAPE.

Dear Mrs. Berry:—   Granite, Okla., 5-26-'13.

We received the shipment of eggs in fine shape, well packed, and none were broken. So far we are well pleased. Have them all under hens, and will report the hatch later on. Many thanks.   Respectfully,   M. A. JOHNSON.

HOUDAN HEN VERY SATISFACTORY.

Dear Friend:—   La Grange, Ind., 5-20-'13.

I just want to let you know that the Houdan hens came yesterday in fine shape, and I am well pleased with them. I am thanking you very much for them. If I ever send for anything else in the poultry line, I will give you a call.
                                 Yours truly,   MRS. SAMUEL BORNTRAGER.

PROUD OF HIS INVESTMENT.

Dear Mrs. Berry:—   Legal, Okla., 10-18-'12.

Received my incubator and brooder yesterday, and everything arrived all O. K. I am sure proud of my investment.   Yours truly,   R. E. McCORD.

BUFF ORPINGTONS.

Orpingtons are one of the newer breeds in the United States, being of English origin. They have sprung into popularity very rapidly in this country. They are large and heavy, but splendid layers. They are splendid foragers, yet on the other hand bear confinement well.

On the Golden Rule Poultry Farm we have raised some of as fine Orpingtons as I ever saw and I can honestly say that they are a fowl that has real merit.

The Orpington is much like the Buff Cochin only do not have the feathers on the shanks, a point in their favor. They are buff in color and buff to the skin, with white skin and white or pinkish shanks. They are in shape like the above illustration, which is from one of our pens and in size between a Barred Rock and the Cochin. They are hardy, fine layers, good mothers, quick broilers, and lots of good points in their favor.

I am proud of our Buff Orpingtons and have a grand lot of them, among them many prize winners. We can care for your orders and assure you we can please you and that you will be delighted with this breed.

Price: Eggs, pen 1, $2.00 per 15; $3.50 per 30; $5.00 per 50; $9.00 per 100; $14.00 per 200. Pen 2: $1.75 per 15; $3.00 per 30; $4.00 per 50; $7.00 per 100; $13.00 per 200.

Price of Stock: Single birds, $3.00; pairs, $5.00; trios, $7.00; pen, 1 male and 4 females, $11.00; pen, 1 male and 10 females, $22.00.

Extra fine lot of fancy cockerels in quantities. Write for special price. See special mating, page 32.

A GOOD HATCH.

Dear Mrs. Berry:— Tarkio, Mo.
I am sorry to have delayed in advising how well the eggs hatched. I want to tell you how remarkably well they hatched. From the 22 duck eggs we have 20 ducks and from the 18 turkey eggs, 12 turkeys. Most sincerely, MRS. M. B. GIFFEN.

WHITE ORPINGTONS—Famous Kellerstrass Crystal Strain.

The most popular chicken at the present time.

The White Orpington is very similar to the Buff Orpington, except in color. They are a very popular fowl on account of their size, laying qualities, hardiness, early maturity, and because they are a fad; but fads are usually something that has merit.

Above picture is from a photo. An idea of their appearance can be obtained.

The Orpington has a very full breast with lots of meat, also very large thighs; the meat is of a very fine grain, juicy and full of flavor.

They are usually ready for broilers about three weeks earlier than many other breeds, as they grow so fast.

The pullets lay early, some claiming that they lay when five to six months old; they will, and ours will excel the 200-egg hen per year, some of them are making a record of 245 each in a year.

These birds bear confinement well and will do well on small quarters. They are splendid foragers and they will pick up nearly half their living when given a chance. They are large, yet splendid layers.

I can furnish you the best strains and something you will be proud to own.

My strain of the White Orpingtons is the best, being direct descendants of the famous Kellerstrass Crystal strain. They have sold for higher prices than any other variety of poultry ever introduced in the history of the world. They have changed hands at thousands of dollars each.

People are buying and raising them because they have real merit. They have the quality to make good in every way, but especially in the production of eggs. They fill the egg basket and that is what brings in the cash commercially; but

WHITE ORPINGTONS—(Continued)

there are few if any eggs put on the commercial market as yet; although there are thousands and thousands, the eggs are all used for hatching purposes during the breeding season and in most cases this season runs longer than any of the other varieties. The White Orpington probably has more good qualities than any other fowl yet discovered.

Get in the White Orpington procession; raise something that your neighbors will all be wanting to purchase a setting of eggs to get the start and will give you more than they would for any other kind.

Mr. Kellerstrass, who lives near Kansas City, Mo., has the largest White Orpington Farm in the world. He sold over $40,000 worth of pure bred eggs and stock, his profits are something enormous; you may not be able to go at it so extensively, but you can make a success of the White Orpington business.

I can start you right with birds and eggs from the best strains. I have one mating from Cook, who has been importing them for several years as a very fine flock. One of our farms raised 460 White Orpingtons this year and they were beauties. Send in your orders early, as I did not have near enough to meet the demand last year and had to disappoint a good many by returning their money.

I was just in time to place here for your information that in strong competition at the Iowa and Southwestern Poultry Show I had the best and highest scoring pen. This shows we have White Orpingtons of quality, and you should have some.

Price of Eggs: Pen No. 1, $2.50 per 15; $4.50 per 30; $7.50 per 50; $14.00 per 100. Pen No. 2, $2.00 per 15; $3.75 per 30; $6.00 per 50; $11.00 per 100; $18.00 per 200. Pen No. 3, $1.75 per 15; $3.25 per 30; $5.00 per 50; $9.00 per 100; $15.00 per 200.

Price of Stock: Extra select—single, $3.50; pairs, $6.00; trios, $10.00; 1 male and 4 females, $14.50; 1 male and 10 females, $30.00. Utility—single, $3.00; pairs, $5.00; trios, $7.00; 1 male and 4 females, $12.00; 1 male and 10 females, $25.00.

RHODE ISLAND REDS—Single and Rose Comb.

The above variety no longer continues to be a new breed among the American people. It has won thousands of admirers, no one turns them down or has an ill word to say against them. A great many class them with the Barred Rock, while others think that they are superior; this in itself shows the interest that must be taken in them.

Each year adds an additional number of admirers. There is no question but that their rapid strides into popularity and general use is not without appreciation of real worth of this quick maturity, hardy, great egg producers.

They are not admired wholly on account of their usefulness as direct money makers; but they are a very beautiful bird; the males of a rich, red color and has a very erect carriage. The females are of a dark buff to a red; they lay when quite young.

The Rhode Island Reds have come to stay and they are certainly worth the place they have won. They are unsurpassed as a table fowl and take the lead as egg producers in the general utility class.

We have a splendid lot of the best Reds I know of and I do not think there are better any place. We started with the best we could obtain and have bred up to the highest standard of perfection and greatest laying records. You will find that I have priced them low compared to the quality.

I have four pens each of the single and rose comb, and can supply the demand with strictly fertile, fresh eggs. You simply cannot lose to buy either stock or eggs of this variety.

Price: Eggs, pen 1 and 2, $2.00 per 15; $3.50 per 30; $4.50 per 50; $7.00 per 100; $13.00 per 200. Pens 3 and 4: $1.75 per 15; $3.00 per 30; $3.50 per 50; $6.00 per 100; $11.00 per 200.

Price of Stock: Single birds, $3.00; pairs, $5.00; trios, $7.00; pen, 1 male and 4 females, $11.00; pen, 1 male and 10 females, $23.00. See special matings, page 32.

Write us for Special Prices on cockerels of any breed in quantities.

LIGHT BRAHMAS.

This breed is the oldest of all pure bred fowls and has stood the rivalry well and are very popular. They are the heaviest weight of all fowls. They have massive heads; hackles white; each feather having a distinct black stripe extending the whole length. The breast, back and under side of the body are pure white on the surface, the lead color under the surface not being considered a defect. The wing bows are white and the primaries of the wings a mottled black and white, the white predominating. The tail is black, the secondaries being laced with white.

The saddle feathers in the males have black stripes in them usually, although the Standard makes no distinction when they are pure white.

Light Brahmas are sturdy, hardy and endure severe weather perfectly. They have very low pea combs and do not suffer from freezing as easily as the breeds with larger combs.

They are gentle, handsome, good layers, good large bodies and are especially recommended for those in towns and cities where but little space can be devoted to chickens. They bear confinement well.

I have a fine lot of these and we are sure that Berry's Golden Rule Farm Brahmas will please you in every way and you will do well if you get some stock or eggs from us.

Price: Eggs, $1.75 per 15; $3.00 per 30; $4.00 per 50; $7.00 per 100; $12.00 per 200.

Price of Stock: Single birds, $3.00; pairs, $5.00; trios, $7.00; pen, 1 male and 4 females, $11.00; 1 male and 10 females, $23.00.

We have the best line of cockerels ever offered and at reduced rates.

Come, I want your order.

WELL PLEASED WITH HOUDAN COCKEREL.

Dear Mrs. Berry:— North Bangor, N. Y., 1-10-'13.

Received the Houdan cockerel in fine condition, and am well pleased with him. Will send the crate back in the morning.

Respectfully yours, MRS. H. P. NORTHUP.

BLACK LANGSHANS.

The above illustration is a true representation of our Black Langshans. They are a distinct breed and not made up from several breeds and is one of the oldest types of fowls. It is claimed to have had its origin in Northern China, the first specimens having been imported to this country about 25 years ago. Since this time they have gained popularity so rapidly that today Lanshans are one of the mainstays in the poultry world.

In plumage they are a rich, glossy black throughout, showing a green lustre. The flesh is white in color, is fine grained, juicy and delicately flavored.

The Langshans are claimed to be very good winter layers and from our experience with them we have found this claim correct. Black Langshan chicks, when hatched, are about one-half white, but at maturity they will be solid black. It often occurs that the lightest colored chick makes the best colored bird when matured. Do not fail to give these lordly birds a trial.

We can please you as to quality and supply you something with which you can lay it over the other fellow at the poultry shows next fall.

Price: Eggs, $1.50 per 15; $2.75 per 30; $3.50 per 50; $6.00 per 100; $11.00 per 200.

Price of Stock: Single birds, $3.00; pairs, $5.00; trios, $7.00; pen, 1 male and 4 females, $11.00; pen, 1 male and 10 females, $22.00.

High class cockerels; have some extra choice ones to offer at special prices in quantities. See special matings, also page 32.

WELL PLEASED WITH HATCH.

Dear Madam:—   Elwood, Nebr., 5-19-'13.

Received the Barred Plymouth Rock eggs alright, the 26th. I placed them under a hen the 28th, and today, May 19th, I took off 37 chicks. I am very well pleased.

Respectfully yours,   J. P. HANLIN.

PARTRIDGE ROCKS ARRIVE IN GOOD CONDITION.

Dear Madam:—   McCleary, Wash., 2-25-'13.

The trio of Partridge Rocks arrived in good condition. They look like fine birds, and hope they will start laying very soon.

Very truly,   B. E. Fleming.

WHITE LANGSHANS.

The above cut shows our White Langshans. They are a new variety that have given most excellent satisfaction, and are more popular than the Black Langshans. Some of our friends who are raising them say that they lay the year around and are a most elegant table fowl. Our record on the Golden Rule Farm bears them out in this.

We have some extra pure birds to sell and a fine flock to furnish eggs from.

They resemble the Black Langshans in every way except that they are white instead of black, and this is certainly a great improvement. I am very sure that you will like them and I advise you to get a start.

Price: Eggs, $1.75 per 15; $3.00 per 30; $4.00 per 50; $7.00 per 100; $12.00 per 200.

Price of Stock: Single birds, $3.00; pairs, $5.00; trios, $7.00; pen, 1 male and 4 females, $11.00; pen, 1 male and 10 females, $23.00.

WHITE LANGSHAN COCKERELS, VERY FINE.

Washta, Ia.
Mrs. A. A. Berry,
Clarinda, Ia.

Dear Mrs. Berry: We received the four White Langshan cockerels all O. K. We were well pleased with them, they are certainly very fine.

Respectfully,
MRS. C. B. BUSH.

Mrs. C. B. Bush, Washta, Ia., feeding her fine flock of BERRY Strain of White Langshans. Aren't they beauties? They are great favorites.

Copyrighted 1905

BUFF COCHINS.

This is a grand old breed of heavy weights. They are very docile and gentle, bearing confinement well. For those situated with little room for poultry, we recommend this breed. They are a town and city man's breed.

They can often be kept in a pen, the fence of which is not over two or three feet high and if you wish to live in peace with your neighbors, get some of these chickens. They are excellent winter layers. They are a solid rich buff color, large, massive, striking in appearance. Legs heavily feathered to the toes.

Continuous breeding for past fifty years along one line has resulted in a breed that is true to type and with their hardiness, docility and adaptation to small breeders, they are immensely popular.

I have them of the best and the Golden Rule Farm has some prize winners and at prices that cannot be duplicated when the high quality is considered. Our trade is very good on this variety.

Price: Eggs, pen 1, $1.75 per 15; $3.00 per 30; $4.00 per 50; $7.00 per 100. Pen No. 2: $1.50 per 15; $2.50 per 30; $3.50 per 50; $6.00 per 100.

Price of Stock: Single birds, $3.00; pairs, $5.00; trios, $7.00; pen, 1 male and 4 females, $11.00; 1 male and 10 females, $23.00.

BUFF COCHIN BANTAMS.

Probably the most popular strain of bantams are the Buff Cochins, which are pretty and profitable pets, as they are so tame and docile. They are a little larger than most varieties of bantams, being broader and heavier set.

All who see them admire them for the unique appearance and beauty. They are quite good layers of rather small eggs, light in color, are good setters and good mothers. Chicks are easily raised, are hardy, both as fowls and chicks. Bear confinement well, thrive as well in small enclosures as running at large; fine pets to have on a nice green lawn, where they command the admiration of all who see them.

Our breeding stock is fine. None better.

Price: Eggs, $1.75 per 15; $3.00 per 30.

Price of Stock: Single birds, $2.00; pairs, $4.00; trios, $5.75; 1 male, 4 females, $9.00.

PARTRIDGE COCHINS.

Partridge Cochins are considered one of the most popular varieties of the Asiatic family. They are certainly a most beautiful fowl and are bred very extensively throughout the entire country. The illustration on this page, by one of the best known artists, is certainly, a life-like reproduction of this beautiful bird. Partridge Cochins are excellent winter layers and if kept in comfortable quarters will lay during the entire winter. They are good for table purposes. Their eggs are large, of rich reddish brown color, and are considered by those who like brown eggs as the best for table purposes. The stock of Partridge Cochins that we breed are prize winners.

We have a large and vigorous flock of this popular variety and can furnish stock or eggs to our customers that we can guarantee in every particular.

Price: Eggs, $2.00 per 15; $3.50 per 30; $5.00 per 50; $10.00 per 100.

Price of Stock: Single birds, $3.00; pairs, $5.00; trios, $7.00; pen, 1 male and 4 females, $11.50; 1 male and 10 females, $24.00.

THE CHICKENS WERE FINE.

Notice the little fellow in the picture; don't he look interested? He is interested, he is the junior member of a firm that have been good customers of ours for a long time. It will not be a great while before he can send me an order. He will know where to get good stock, he will buy where his father and mother used to send for their birds and supplies.

WOULD NOT TAKE $15.00 FOR HIS CROWER.

Statesville, N. C.
Dear Madam:—I received my crower today, and I would not take $15.00 for him, I returned the coop to the express office today. J. M. SUTHER.

SINGLE COMB BROWN LEGHORNS.

This variety is so well known that we need not enter into a detailed description of them. They are of the Mediterranean class of fowls and consequently non-setters. They are known the world over as the best layers we have in the feathered tribe, the hens often lay during the time they are in moult.

The eggs are of a pure chalky white and of medium size. The male is very beautiful, having orange hackle and saddle, with black stripes in center of each feather; ear lobes are pure white. The female is a medium brown with delicate penciling on back, and wings are of a darker brown, hackle similar to that of the male; breast salmon colored.

I have a fine flock of Single Comb Brown Leghorns that are line bred to obtain the greatest efficiency in laying. If you desire as fine birds as ever graced a poultry yard, do not fail to send to us.

Price of Pen 1: Eggs, $1.50 per 15; $2.50 per 30; $4.00 per 50; $7.00 per 100; $12.00 per 200.

Pen 2: $1.25 per 15; $2.00 per 30; $3.50 per 50; $6.00 per 100; $10.00 per 200.

Price of Stock: Single birds, $2.50; pairs, $4.50; trios, male and two females $6.50; pen, 1 male and 4 females, $10.00; pen, 1 male and 10 females, $20.00.

A fine lot of cockerels to offer at special prices in quantities.

WELL PLEASED WITH PARTRIDGE COCHIN COCKEREL.

Dear Madam:—   Kansas City, Mo., 3-5-'13.

I received the Partridge Cochin cockerel yesterday, and am well pleased with it. As I live outside of the delivery limits I went to the office to get it. I scratched out my name on the coop as directed by you, and ordered it sent back to Clarinda. Hoping it will reach you in good shape, I am.

                              Yours truly,   I. McNELLIS.

WHITE ROCKS RECEIVED IN GOOD SHAPE.

Dear Madam:—   Marquette, Kas., 1-13-'13.

I received the pen of White Rocks in good shape. Will send the crate back soon.

            Yours truly,   ROY C. MILLER.

ROSE COMB BROWN LEGHORNS.

This class of the Leghorn family is like the Single Comb Brown Leghorn in every respect except that this variety has rose or pea combs, a qualification that is very desirable, as they are not bothered with any frozen combs or wattles, and for northern sections of this country this advantage is of considerable value.

For a handsome bird and for egg producers the Leghorns stand at the head of all varieties. They produce more eggs with less feed than others. They are very hardy, active, great hustlers and require plenty of room. For pure plumage, correct shape and form, with great egg producing qualities, our Rose Comb Brown Leghorns stand at the head.

Price of Pen 1: Eggs, $1.50 per 15; $2.50 per 30; $4.00 per 50; $7.00 per 100; $12.00 per 200.

Price of Pen 2: Eggs, $1.25 per 15; $2.00 per 30; $3.50 per 50; $6.00 per 100; $10.00 per 200.

Price of Stock: Single birds, $2.50; pairs, $4.50; trios, $6.50; pen, 1 male and 4 females, $10.00; 1 male and 10 females, $20.00.

We have a fine lot of cockerels that we will take pleasure in making special price in quantities. We can please you.

BARRED ROCK ARRIVES IN FINE CONDITION.

Dear Madam:— Stockton, Cal., 3-11-'13.

The Barred Plymouth Rock cockerel arrived in fine condition and am well pleased with same. Returned the crate by Wells/Fargo Express today. Thanking you again, I am.
Yours truly, H. A. JANDEBEUR.

PROUD OF THE BUFF ORPINGTON COCKEREL.

Dear Madam:— Branson, Mo., 3-17-'13.

A short time ago you shipped me a Buff Orpington cockerel of which I am very proud. I wish to order some eggs and perhaps some fowls this summer, and wish you would send me your latest catalog.
Very truly yours, MRS. RUSSELL BROWN.

NAMED HIM WOODROW.

Dear Madam:— Waterloo, Neb., 3-6-'13.

The Plymouth Rock cockerel you shipped to us arrived in good condition yesterday. We think he is a fine bird and have named him Woodrow. We ordered the crate returned at the station. Yours truly, R. W. BARBER.

WHITE ROCK COCKEREL CERTAINLY A FINE BIRD.

Dear Madam:— St. Paul, Minn., 3-24-'13.

Received the White Rock cockerel all O. K. and am well pleased with him. Returned crate the 20th. You probably have received it by now. Will certainly remember you for future business. I beg to remain, Your friend, O. S. BROWN.

SINGLE COMB WHITE LEGHORNS.

This breed is very much like the Brown Leghorn except in color which is pure snow white. It appears, from the present demand for the white variety, that they may give their cousins a close race for first place.

The White Leghorns are being used a great deal by breeders who cater exclusively to egg trade, as it is claimed that they lay a slightly larger egg than the Brown Leghorns and just as many of them.

They have a yellow beak and shank and have received some attention from the broilermen on account of rapid growth and because of their small size they may be marketed profitably as squab broilers. From a beauty standpoint they stand right in the front rank.

We have a fine lot from which to fill our orders and will not disappoint you.

Price of Pen 1: Eggs, $1.50 per 15; $2.50 per 30; $4.00 per 50; $7.00 per 100; $12.00 per 200.

Price of Pen 2: Eggs, $1.25 per 15; $2.00 per 30; $3.50 per 50; $6.00 per 100; $10.00 per 200.

Price of Stock: Single birds, $3.00; pairs, $5.00; trios, $6.50; pen, 1 male and 4 females, $10.00; 1 male and 10 females, $21.00.

PROUD OF THE BUFF ROCKS.

Dear Madam:— San Angelo, Tex., 3-7-'13.
    The Buff Rocks I bought of you arrived yesterday and in good shape. They are dandies and I am proud of them. I had the coop sent back the 6th. Thanking you for quick delivery, and wishing you continued success, I am. F. P. NIPP.

RESULT IN ADJUSTING CLAIM SATISFACTORY.

Dear Madam:— Magnolia, Ark., 11-'3-'13.
    The cockerel arrived this A. M. and must say I am well pleased with him. Many thanks for your adjustment of my claim. I will do all I can for you in this part of the country. Yours truly, W. P. LONGINO.

ROSE COMB WHITE LEGHORNS.

This is one of the popular breeds of the heavy laying varieties, and are identical with the Single Comb White Leghorns, except that the comb resembles the combs of the Hamburg, and are sometimes called pea-comb Leghorns. They are much admired by poultry fanciers and those having egg farms, being wonderful egg producers.

Their freedom from frozen combs makes them more desirable for our northern climate than the single comb varieties. They are very stylish and make a very attractive appearance. They belong to the non-setting class of fowls, and are unexcelled as layers by any of the Leghorn or other families. The eggs are very white in color and medium size. They are very hardy both as chicks and fowls. Chicks grow very fast. Pullets frequently begin laying at four months old. They are a very fine fowl for the table as far as they go, but being rather small, we could not recommend them as a valuable market fowl, but so far as turning cents into dollars in the production of eggs, they stand in the front rank. We have a very nice flock of choice birds, with a great laying record, and we strongly recommend them to any one looking for this breed.

Price of Pen 1: Eggs, $1.50 per 15; $2.50 per 30; $4.00 per 50; $7.00 per 100; $12.00 per 200.

Price of Pen 2: Eggs, $1.25 per 15; $2.00 per 30; $3.50 per 50; $6.00 per 100; $10.00 per 200.

Price of Stock: Single birds, $3.00; pairs, $5.00; trios, $7.00; pen, 1 male and 4 females, $11.00; pen, 1 male and 10 females, $22.00.

BIRDS SATISFACTORY.

Dear Madam:—   Milwaukee, Wis., 2-11-'13.

I received the chickens in good condition, and I am well satisfied. Thanking you very much for the attention given my order, I am.   Yours truly,   FRANK HELIMESKI.

WELL PLEASED WITH PEN OF LEGHORNS.

Dear Madam:—   Escanaba, Mich., 10-14-'13.

Received the pen of White Leghorns O. K. and was well pleased with the birds. Excuse me for not writing you sooner, but have been busy house-cleaning and did not have the time.
                       As ever,   MRS. J. M. UTT.

Order your stock and eggs from an experienced breeder, one who understands the mating of thoroughbred fowls to produce good results.

SINGLE COMB BUFF LEGHORNS.

This variety of the Leghorn family is comparatively new and differs but little from their cousins except in color, which is a splendid buff, so much admired in poultry. They are becoming very popular and justly so, as they are a most excellent breed. Are good layers, non-setters, rapid growers and possess the yellow skin and shank. As is the case with other Leghorns, they are high flyers and will fly from enclosures if not made sufficiently high or if not covered. They make excellent foragers if left to roam, as they are naturally active and should be kept busy if confined.

We wish to say that we have a fine flock of Buff Leghorns from which to fill our orders and can promise something extra.

Price of Eggs: $1.50 per 15; $2.50 per 30; $4.00 per 50; $7.00 per 100; $12.00 per 200.

Price of Stock: Single birds, $2.50; pairs, $5.00; trios, $6.50; pen, 1 male and 4 females, $11.00; 1 male and 10 females, $22.00.

Have a large supply of extra fine cockerels to offer at a bargain in quantities.

WILL BE REMEMBERED WHEN IN NEED OF ANYTHING FURTHER IN OUR LINE.

Dear Madam:— New Orleans, La., 11-6-'13.
The Light Brahmas received O. K. Thanks for such a good selection. Also want to thank you for the cockerel you sent with the pullets. He will make a fine cock-bird. Wishing you all kinds of success in your business, and will remember you again when in need of anything in your line of business. I am, J. C. RICHARDS.

BLACK MINORCAS.

Single and Rose Comb.

The Black Minorca is a well established breed of English fowls belonging to the Spanish varieties and, wherever bred, are considered a valuable breed, and hardy both as fowls and chicks, easily raised, mature early, and pullets commence laying very young.

Is one of the most stately and aristocratic of all breeds recognized by the standard, and at the same time a very profitable breed to keep.

This variety is black from the tip of its beak to the end of its toes, but when we say black we do not express it all, for the plumage of the cock fairly glitters with a brilliant green irridenscence quite impossible to describe.

The comb of the cock is very large and upright, divided into six spikes and extending back over the head. The wattles are red, long and pendulous, the ear lobes pure white and of the texture and appearance of the finest kid leather.

They are non-setters, small eaters, splendid foragers, and without doubt very profitable. Their adaptability to all soils and places, whether in confinement or allowed unlimited range, makes them very popular, and suitable to the city fancier as well as the farmer. Their plumage is a pure black with a green or metallic lustre. Their legs are nice and smooth and medium length. The chief and striking ornament of the cock is his comb, which is very large, single, straight as an arrow and even serrated; has a large flowing tail, carried somewhat high. My stock of this variety is simply first-class, the best I could get, regardless of price.

We have in our breeding pens birds that will score high and that are good layers, and our friends will find our stock to be of the kind that wins at shows. We can supply eggs from as good stock as this country affords.

Price: Eggs, $1.50 per 15; $2.50 per 30; $4.00 per 50; $7.00 per 100; $12.00 per 200.

Price of Stock: Single birds, $3.00; pairs, $5.00; trios, $7.00; pen, 1 male and 4 females, $11.00; pen, 1 male and 10 females, $23.00.

WHITE MINORCAS—Single and Rose Comb.

The White Minorcas originated in Spain and are a very old variety. How they originated is not known; but the supposition is that they are sports from the Black Minorcas, as it is a well known fact that a black fowl will occasionally throw off a white chick.

The White Minorcas are similar to the Black in every way with the exception of color, and as many prefer the white chicken they are very popular.

They lay a very large chalk white egg and are splendid layers; they are larger than the Leghorn family and are a better table fowl, in fact they rank very high as a table fowl; while great in egg-production, the hens laying from 200 to 250 eggs in a year.

As to vitality and productiveness and quality they equal their cousins, the Black Minorcas.

They are the same build, have coral red faces, white ear lobes and of same size, they are non-setters, great foragers and rustlers.

They have splendid plumage, very gay and attractive in appearance, having the true Minorca shape and good qualities in general.

Now I have a fine flock of these and can offer you an extra big value for your money. If you want something extra good in egg production and an extra large fancy egg and still good table fowl get some Minorcas, and as to color that is a matter of choice.

Price: Eggs, $2.25 per 15; $4.00 per 30; $6.50 per 50; $10.00 per 100.
Price of Stock: Single birds, $3.00; pairs, $6.00; trios, $8.00; pen, 1 male and 4 females, $13.00.

HOUDANS

Originally from France, a very old and popular strain of thoroughbred fowl. They are black and white, the white showing in spots as the above cut, which is a good representation of this breed.

They are splendid layers and produce a large white egg. They are non-setters and in our years of experience very few have gone to setting. They are probably the largest breed of heavy layers and non-setters.

The Houdans have five toes, a crest and beard, and are shaped much like the Dorkings, with the long body, and are noted as a table fowl, meat tender, juicy and fine flavored.

The chicks are beauties, grow very fast, often weigh 4 to 5 pounds at four or five months of age. They are very hardy and thrive under ordinary care. They are especially adapted for a city or town, where quarters are limited, and they are gentle and docile and will not fly away or roam far away. They make splendid pets.

The pullets commence to lay early and are productive to greater age than most varieties, as they will produce eggs profitably at an age of 4 or 5 years. They are a small eater of most any kind of wholesome food. Splendid winter layers, no better. They are a good sized chicken and are what could be called general utility. Standard weight, cock, 7 lbs., and hen, 6 lbs.

We have been breeding them for years and started with the best and have used all the skill and knowledge we possess in improving them.

I will say that they are receiving more attention from lovers of chickens every year, and becoming one of the most popular varieties. They have real merit.

We are offering you the best at a price than cannot be matched by smaller breeders when quality is considered.

Price of Eggs: $2.00 per 15; $3.25 per 30; $4.75 per 50; $8.00 per 100.

Price of Stock: Single birds, $3.00; pairs, $5.00; trios, $7.00; pen, 1 male and 4 females, $11.00; pen, 1 male and 10 females, $22.00.

Plenty of choice high grade cockerels at special prices in quantities.

First class guaranteed birds at $2.00 each.

HOUDANS START LAYING ONE WEEK AFTER ARRIVAL.

Dear Madam:— Hansboro, N. Dak., 6-9-'13.
Am in receipt of the pen of French Houdans, and am well pleased with them. They stood shipment well, and started laying one week after arrival. Returned the crate the day after I received the birds. Yours very truly, B. L. THOMAS.

CORNISH INDIAN GAMES.

The Cornish Indian Game stands in the very forefront as a table fowl, its massive breast meat, thick, solid thighs and large size, all made up of fine grained, sweet and juicy flesh, make it the ideal fowl for the table.

As layers the Indian Games are only fairly good, their eggs being of good size, however, and of fine flavor. They are good but not persistent setters, and as mothers cannot be excelled by any breed. They have a very high carriage and bold bearing, being perfectly capable of taking care of themselves.

In color they are a deep mahogany, each feather penciled with two lighter brown stripes.

They are among the hardiest fowls, being covered with a thick and very close coat of feathers, and have very small pea combs. They endure very severe weather without becoming disfigured by frost. Give them a warm house to sleep in and feed them properly, and they will lay eggs in winter quite freely.

When I concluded to add Indian Games to our list we carefully selected stock from the best strains, and have bred it so as to be able to mate up our pens without breeding related stock together.

The result is that our pens have in them a choice lot of very vigorous birds, from which fine specimens may confidently be expected.

The above illustration will give those unacquainted with this grand breed a better idea of the shape, marking, stately bearing and general appearance of our strain of Cornish Indian Games than could be given by pages of descriptive matter.

For breeding fowls for market purposes, or for improving common stock by cross breeding, no breed is better than the Cornish Indian Games.

If you want to get these from headquarters, send to me.

Price: Eggs, $2.00 per 15; $3.50 per 30; $4.50 per 50; $7.50 per 100.
Price of Stock: Single birds, $3.00; pairs, $5.00; trios, $7.00; pen, 1 male and 4 females, $11.00; 1 male and 10 females, $23.00.

"CORNISH INDIAN GAMES GOOD IN EVERY RESPECT," IS THE VERDICT OF A PLEASED CUSTOMER.

Model Steam Laundry, Atlantic, Ia.

Dear Madam:—I received the two Cornish Indian Games this morning and must say that I am well pleased with the birds. They are good in every respect and I have nothing but praise for your concern and will place all my orders with you in the future.

Very truly yours, A. C. JENSEN.

MOTTLED ANCONAS.

The Anconas are a breed not so well known throughout the United States, although they are gaining in popularity. They originated from the Minorcas, hence they have all the good qualities of the Minorcas. They are in many respects like the Leghorn, but larger. We are safe in saying that they are as good layers as any breed in existence. They lay a nice white egg of good size. We do not even express ourselves in saying that they are an "Egg Machine," so great layers are they. They are also non-setters making them more popular in the production of eggs. They are small eaters, splendid foragers, and without doubt very profitable. Their general color is black with white feathers mixed through the black giving them a spotted appearance. They are fine lookers and you will be proud to own a flock of them. The above cut is a good representation of this breed. We can highly recommend this breed as egg producers. This breed of fowls is very attractive and no one can help but like them. We have a fine flock of these. No better any place, and you can get the best at the Berry Poultry Farm. We have spared neither time nor money in production of as fine a flock of these as there is in the country, and they are so very satisfactory that we recommend them to any one wanting something fine, and with great laying qualities.

Price: Eggs, $2.00 per 15; $3.50 per 30; $4.50 per 50; $7.50 per 100; $12.00 per 200.

Price of Stock: Single birds, $3.00; pairs, $5.00; trios, $7.00; pen, 1 male and 4 females, $11.00; pen, 1 male and 10 females, $23.00.

A fine lot of choice cockerels, write for special prices in quantities.

CHICKENS RECEIVED IN GOOD CONDITION.

Dear Mrs. Berry:— Hemingford, Nebr., 2-10-'13.
The chickens you sent me arrived in good condition the 5th inst. and I thank you for the promptness in filling the order. I returned the crate the 8th inst. Hope you receive it all O. K. Yours truly, MRS. C. M. LOTSPEICH.

Always keep a supply of Berry's insect powder and lice killer on hands all the time; the best disinfectants and vermin destroyers on the market.

SILVER SPANGLED HAMBURG.
Are of the Mediterranean Class or Egg Producers.

The Silver Spangled Hamburgs are one of the most beautiful varieties that can be found on the poultry list. No one can pass a flock of them without a glance of admiration. For beauty they are unsurpassed, and too much could not be said of this beautiful breed. As egg producers they stand in the front rank, laying the year around. In color eggs are white, and medium size. Chicks grow quite fast and mature very early. Pullets often begin laying at four months old. They are very small feeders, and bear confinement in small enclosures remarkably well.

For laying qualities and beauty they stand on their own merits and cannot be overestimated. Hens will weigh from 4 to 5 lbs.; cocks from 5 to 6½ lbs.

The Hamburgs are an excellent breed, being very prolific, so marked being this characteristic that they were at one time called "Dutch Everlasting Layers." The Silver Spangled Hamburg is admitted to be the most beautiful of the Hamburg varieties. It is clear white and black, the two colors forming a striking contrast to each other. The black is disposed in round spots, the end of each feather being one of these round black spots. In good specimens these spots are regularly set in rows, making the breed distinctly an ornamental one, besides its value in egg production.

Price of Eggs: $1.75 per 15; $3.50 per 30; $4.50 per 50; $7.00 per 100; $12.00 per 200.

Price of Stock: Single birds, $3.00; pairs, $5.00; trios, $7.00; pen, 1 male and 4 females, $11.00; pen, 1 male and 10 females, $22.00.

A fine lot fancy cockerels at bargain prices.

Introduce new blood into your flocks every year. Order your eggs from Mrs. A. A. Berry.

THANKS US FOR PROMPT ATTENTION AND FINE BIRDS.

Mason City, Ia.

Dear Friend:—We received the Hamburgs in good condition yesterday, and they are fine. I do certainly thank you for the prompt attention and sending them so soon. We want a couple of settings of eggs just as soon as the hens commence to set.

We will send the crate back soon.

Yours truly,

MRS. F. D. WAY.

Mrs. F. D. Way and her son among their pets. A fine flock of hamburgs.

WHITE FACED BLACK SPANISH.

This very popular and handsome breed belongs to the Mediterranean class, hence are classed with the non-setting variety. Their combs and wattles are bright red and the face being pure white, they present such a striking appearance that makes you proud to own them, and are admired by all who see them. They are undoubtedly handsome.

Their plumage is a nice glossy black throughout. They are great layers, producing mostly large white eggs which usually hatch well, as the Spanish possess great vitality. These birds mature very fast, and pullets begin laying while yet quite young and are good winter layers. They bear confinement quite well and are good foragers if left to roam. They are a good table fowl, having sweet tender meat that is very juicy. They weigh about the same as the Black Minorca; in fact, they are about the same size. Cocks weigh from 6 to 8 lbs.; hens from 5 to 7 lbs. We have bred them for several years and find them to breed very true to color, and splendid fowl for egg production and trade in general.

Price: Eggs, $2.00 per 15; $4.00 per 30; $5.00 per 50; $8.00 per 100.

Price of Stock: Single birds, $3.00; pairs, $6.00; trios, $8.00; pen, 1 male and 4 females, $12.00.

DUCKS MAKE GOOD IN LAYING.

Dear Mrs. Berry:—   Hart, Mich., 10-3-'13.

The ducks were received in good condition. Was well pleased with them. My little grandson said they were "dandies." Today he reports that one of the ducks laid an egg the 30th of September, and that she had laid one every day since.

Respectfully,   MRS. F. D. SCHLAPPI.

THINKS BARRED ROCK WILL MAKE A FINE BIRD.

Dear Madam:—   Wharton, N. J., 11-4-'13.

I received the Barred Plymouth Rock in good condition. I am well pleased with him, and we think he will make a fine bird. Thanking you for your promptness in filling my order, and with best respects, I am,

Yours truly,   G. T. WATERS.

WELL PLEASED WITH PEN OF PARTRIDGE WYANDOTTES.

Dear Madam:—   Gasport, N. Y., 12-6-'12.

I received the trio of Partridge Wyandottes and am well pleased with them. I think they will be very large and healthy, when they get their full growth. They are very healthy now. I will send the coop back as soon as the team goes to the express office.

Yours truly,   GEORGE CHAPMAN.

THE BLUE ANDALUSIANS.

This handsome breed of fowls is of English origin, dating back over forty years, when they had been bred quite extensively in Andalusia, Spain, from whence they derive their name. They belong to the non-setting class. Having bred them for some time, I have never known one to get the least broody. These birds are fine layers both winter and summer, hard to beat; they lay a medium sized egg.

They are very hardy and mature fast and pullets lay while yet very young. Their face comb and wattles are red, with white ear lobes. They are a very attractive fowl, having a haughty appearance and are about the same size as the Minorca. However, they are not without a fault; they do not breed as true to color as most varieties, but their good qualities make up for this weak point.

Our stock is fine and as good as any in the land.
Price: Eggs, $2.50 per 15; $4.00 per 30; $6.00 per 50; $10.00 per 100.
Price of Stock: Single birds, $3.50; pairs, $6.50; trios, $9.00.

WHITE LANGSHANS ALWAYS GIVE SATISFACTION.
Dear Mrs. Berry:— Schoeneck, Pa., 2-6-'13.
I received the pair of White Langshans on the first of March, and I am very proud of them. They arrived in good condition and did not seem to be effected by the cold. What will you charge for two more hens?
Yours truly, KATIE HARTING.

WELL PLEASED WITH BLUE ANDALUSIAN COCKEREL.
Dear Madam:— Jeffersonville, Ind., 1-20-'13.
We received the Blue Andalusian cockerel all O. K. today and I am well pleased with him. Will return the coop tomorrow.
Yours truly, ALEXANDER WENDEL.

VERY PROUD OF BIRDS RECEIVED.
Kind Friend:— N. Middletown, Ky., 1-12-'13.
I am proud to say that my chickens have arrived, and I am well pleased with them. I will return your coop at once.
Yours truly, MRS. NEWTON STRAWDER.

WHITE CRESTED BLACK POLISH.

The above is a good illustration of this most beautiful of all fowls. Jet black in color except the crest which is snow white all but a little just above the bill.

For a fancy fowl, something that will delight the eye and an ornament to any yard, the White Crested Black Polish undoubtedly takes the lead.

This bird is very domestic being extremely tame and docile, they bear confinement well and are splendid fowls for the city and town with small quarters. They are splendid layers and lay a medium sized egg, eggs are usually very fertile and hatch out strong, pretty little chicks.

We raised some nice ones last year at the Golden Rule Poultry Farm and have some fine ones to offer and can supply eggs from a very fine flock. If you want something useful as well as ornamental, get a start of our White Crested Black Polish.

Price of Eggs: $2.50 per 15; $4.00 per 30; $5.50 per 50.
Price of Stock: Single birds, $3.50; pairs, $6.50; trios, $9.00.

PEARL GUINEA FOWLS.

Guineas are of somewhat wild disposition, and the more they run wild the better the quality of the flesh, which is equal to pheasants. The eggs are very much sought after for culinary purposes for fancy cookery. Guineas are becoming very profitable to raise. They cost little, the only requirement being plenty of range. The eggs are very fertile; chicks hardy and easy to raise. Since game has become too scarce a substitute for pheasant, grouse and prairie chickens has been in the Guinea fowls and this industry is receiving much attention by those dealing in game products. We have some nice pure birds. They do not commence to lay until May.

Price: Eggs, $1.75 per 15; $3.50 per 30; $5.00 per 50.
Price of Stocks: Single, $2.00; pairs, $3.50; trios, $5.00.

INDIAN RUNNER DUCKS.

The Indian Runner Ducks are a comparatively new breed in America. They came from West India, which is the original home of these great egg producing ducks. They are the greatest egg producing fowl in the world.

This specie represents the egg producing strain of ducks, as the Leghorn represents the egg producing strain of chickens. They are a regular egg machine.

December 1st several of our June hatched Indian Runners were laying. They lay all the year around except a short time in August and September when fed for it, as they have been cultivated up to an enormous egg producing capacity. Their eggs sell readily in the market; they hatch well and produce thrifty and vigorous ducklings, which grow quickly into mature specimens. These can be readily fattened when young to four and five pounds when only three months of age, and make a very desirable light weight roaster. They will also produce a fair quantity of feathers, but they hold their greatest popularity as egg producers.

If you want to produce eggs all winter, you can do it readily with some of our Indian Runner Ducks, and do it easier and with less expense in regard to feed and care than hens. They are extremely vigorous and hardy.

They are dry land fowls and do not need running water, a pond or lake, to thrive and do well. All that is necessary is a sufficient amount of clean drinking water. You should obtain a start from our Golden Rule Poultry Farm, heavy egg producing strain of Indian Runner Ducks. I am sure that you will like them and that they will make you lots of money.

A FEW WORDS ABOUT THE DIFFERENT KINDS OF RUNNER DUCKS.

There are three strains of the Indian Runner Duck family, the brown and white, sometimes called English penciled, these are supposed to be the original English strain. The fawn and white called the American Standard, then there is the Pure White, these are also an American strain.

The brown and white and the fawn and white are marked just alike, with, of course, the exception, that the brown takes the place of the fawn and visa versa. The above picture shows the way these two strains are marked, the dark represents either the fawn or the brown. The Pure white, of course has the same characteristics as the strains represented in the picture, that is, same egg laying qualities, same shape, etc. The pure white strain is the last to be introduced, and taken up so well among the breeders that there is an exceptionally large demand for them, making the price of those a little higher. The white is preferred so many times to colored varieties of all kinds.

A basket of our Indian Runner duck eggs. Are they not beauties? Please observe the size and quality.

SOME OF THE QUALITIES OF INDIAN RUNNER DUCKS IN BRIEF.
Expert Testimony From High Authority.

Read what F. D. Coburn, the great farmers' friend, of Kansas, says in his report of the Kansas State Board of Agriculture, September, 1908. This is the extract:

"Indian Runner ducks are becoming a popular breed and this is because of their great capacity for laying eggs. Great egg records are given for every flock of Runner ducks."

Now what do you think of that, coming from a man that has done more for the farmers in Kansas and other states than any other man in the United States, and he speaks the truth about the Indian Runner ducks.

Here is what a poultry expert says:

"Indian Runner ducks are popular with everybody; so there must be something to a fowl that so many admire."

Some Records of Indian Runner Ducks.

This is an article from the pen of one of the best poultry writers and judges in the United States, a man that is authority and one who knows:

"Indian Runners seem to make good everywhere in the world. In every part of America they show the same remarkable prolificacy that brought them into such prominent notice in England. In the Australian laying competitions they are making remarkable laying records both in one-year and two-year tests. One pen of six ducks in these competitions made a record of more than 2,300 eggs in twenty-two consecutive months. I own a small pen that comes as near producing an egg a day for every duck in it as is possible. I really cannot understand how they can produce eggs as they do on what they eat.

"In my opinion Indian Runner ducks will soon be recognized as the distinctive egg fowl of this country. As they are not troubled with insect enemies, do not easily catch cold, are immune from roupe, proof against wet weather and hardy in every way, they are a distinct acquisition to our poultry yards." MILLER PURVIS.

If you are contemplating going into the Runner duck business you should by all means call for one of the "Indian Runner Duck Culture" booklets which I will send upon request. It describes and tells all about the Runners from A to Z.

Price of Eggs: Brown and White—"English" strain—$1.75 per setting of 13; $3.00 per 26; $5.50 per 50; $8.00 per 100. Fawn and White—"American" strain—$2.00 per setting of 13; $3.50 per 26; $6.00 per 50; $10.00 per 100. Pure White—"American" strain—$3.00 per setting of 13; $5.50 per 26; $8.00 per 50; $14.00 per 100.

Price of Stock: Brown and White or Fawn and White, single birds, $3.00; pairs, $5.00; trios, $7.00; pen, 1 drake and 4 ducks, $11.00; pen, 1 drake and 6 ducks, $14.00; pen, 2 drakes and 10 ducks, $23.00. Pure White, single bird, $4.00; pairs, $7.00; trios, $10.00; pen, 1 drake and 4 ducks, $15.00; pen, 1 drake and 6 ducks, $19.50; pen, 2 drakes and 10 ducks, $33.00.

MAMMOTH IMPERIAL PEKIN DUCKS.

The best variety of ducks to raise is best adapted to the market demands. Pekin ducks were imported from China many years ago, but the scientific American fancier have since their importation into the country bred them to such a high degree of perfection, mainly in size and shape, that there remains scarcely a drop of foreign blood in them.

The Pekin ducks are very large, white or creamy white in color, laying from 100 to 150 eggs in a season. They are very hardy, easy to raise and at ten weeks old they are in full feather and often weigh five pounds at that age.

They have have bright yellow legs and bills, are excellent foragers and they need no more water than chickens. It is the opinion of many that water fowls should have running water to thrive well. This is not so. They will thrive remarkably well with plenty of water to drink and no water at all to swim in. However, they enjoy themselves much more if provided with water in which to swim or wash themselves.

Ducks and geese are very profitable to raise. We can offer you something very nice in this line and can please you.

Price of Eggs: $1.50 per 11; $2.50 per 22; $4.00 per 50; $6.00 per 100.

Price of Stock: Single ones, $3.00 each; per pair, $5.00; trio, $7.00; pen, 1 male and 4 females, $10.00.

PEKIN DRAKE A FINE BIRD.

Dear Madam:— Bruxe, Wis., 8-16-'13.

The Pekin drake arrived on the 12th of August and am well pleased with it. He is a fine bird. Will return the crate as soon as we go to town. Thanking you for your kindness, I am. Yours truly, MRS. LOUIS NATER.

HAMBURGS VERY NICE.

Dear Madam:— Volney, Va., 9-15-'13.

I received the trio of Hamburgs all O. K. and think they are very nice birds. I thank you very much. Yours truly, V. E. REEDY.

If you expect to succeed in the poultry business you will have to infuse new blood of non-related stock. We can always supply your wants and of the best and highest quality. Flocks with egg laying records are one of our specialties.

TOULOUSE GEESE.

This breed was imported from Europe and is the best variety, very popular and profitable.

They are very large when four years old and when extra well fattened have weighed 25 to 40 pounds to the pair, but their average weight in laying condition is 25 pounds to the pair.

These geese lay from thirty to forty eggs in a season. They lay very early so be sure and send your order early in the spring. Their eggs hatch well. The goslings are much stronger than common ones. They grow very rapidly, and attain a heavy weight.

The feathers are of the best and they often average half a pound feathers to the picking.

They are small feeders and do not require any feed but pasture except in cold weather. In color geese and ganders are just alike, a uniform grey; breast and under part of body a shade lighter, good disposition, easily fenced, breed at one year old and profitable in every way.

Geese usually commence to lay in February or earlier; so do not put off ordering, but do it at once and have us ship in March, April or May.

I have a fine flock from which to fill your orders and can give you stock or eggs.

Price of Eggs: $1.50 per 7; $2.75 per 14; $4.00 per 21; $5.00 per 30; $8.00 per 50.

Pen of one male and four females, $12.00.

Price of Stock: Single birds, $3.50; pairs, $6.00; trios, $8.00.

Dear Madam:— Green Pond, Ala., 4-2-'13.
I received the pen of 7 Buff Orpingtons, and think they are a fine lot of fowls. I have received one egg already. I will send the coop back as soon as I can get to it.
Yours truly, J. M. OGLESBY.

Read everything that we have to say about the "Biddy" incubators and brooders and be convinced that they are the machines you have been looking for.

If there is anything in the poultry line you want and do not see it in our book, "Profitable Poultry," write us about it.

MAMMOTH BRONZE TURKEYS.

The Mammoth Bronze Turkey may certainly be styled "Mammoth," not only more fitly and deservedly than any other turkeys, but better than any of the various breeds of poultry.

They are the monarchs of poultrydom. The Mammoth Bronze Turkey claims its origin from the wild turkey, but to compare them at the present time with the wild turkey as it always has been and the Mammoth Bronze as we have it bred up today, one would hardly realize that they were once the same. They are so well known that a detailed description of their plumage seems unnecessary. The predominating color of both male and female is a brilliant bronze which glistens in the sun like burnished gold. Turkeys must have their liberty to do well, as confinement is very detrimental to them, they being great roamers, and it seems an impossibility to breed it out of them.

They are great foragers and may be raised very cheaply if given sufficient range. But this is essential, as they will not bear confinement.

We have a nice lot to offer and if you want something extra well bred, we can please you.

Price of Eggs: $3.00 per 9; $5.00 per 18; $7.50 per 28.

Price of Stock: Large toms, $6.00 to $12.00; turkey hens, $5.00 to $10.00; pair, $12.00; trios, $16.00; pen, 1 male and 4 females, $25.00. Size, weight and quality govern prices. Write for special prices.

GOLDEN WYANDOTTE COCKEREL SURE A FINE BIRD.

Dear Madam:—   Dedalia, Mo., 4-17-'13.

We received the Golden Wyandotte cockerel alright, and he is sure a fine bird. He is so tame. Thanking you ever so much for sending me such a fine bird, I am.

Yours truly, MRS. S. C. CAPEN.

BIRD IS CERTAINLY A BEAUTY.

Dear Madam:—   Okeene, Okla., 3-28-'13.

I received my bird all O. K. and he is certainly a beauty. Many thanks for the prompt shipment. Am returning coop. Yours truly, MRS. MAYME MILLER.

HERBERT HIGHLY PLEASED WITH CORNISH INDIAN GAMES.

Dear Friend:—   St. Anthony, Idaho, 3-28-'13.

I received the Cornish Indian Games today, and they are all right. They are very nice birds. They are the nicest ones in town, I believe. When I took them out of the crate found that the hen had laid an egg so I think I will have no difficulty in getting plenty of eggs from her. Will return the crate on next train.

Your true friend, HERBERT McINTYRE.

WHITE HOLLAND TURKEYS.

White Hollands are a very hardy bird and is a variety fast coming into popularity.

The breed came from Germany here; but it is supposed that their original home is in Holland.

In size they are a little smaller than the Bronze, which is so universally known, but still they attain a good size. Ours are larger than the most of the common bronze that are found throughout the country. They are handsome and on account of their hardiness are held in high esteem by all that breed them.

I have a magnificent flock of them of splendid size, fine shape, snow white, excellent foragers and quite docile. They are tamer than the Bronze as a rule.

I know that we can please you in White Holland turkeys as there are no better in the land.

Turkeys do not begin laying until in April; so if you order some eggs with fowls or other articles before that date please let us know whether or not you wish us to hold your order until we can fill at once, or ship other goods and follow up with eggs as soon as possible.

As you know that there is a difference in size and quality and a great fluctuattion of the market price of turkeys, it is impossible for us to make a firm price, so it is best to always write for prices especially on quantites.

Price of Eggs: $3.00 per 9; $5.00 per 18; $7.50 per 27.

Price of Stock: Large toms, $6.00 to $10.00; turkey hens, $5.00 to $8.00 each.

WRITE FOR SPECIAL PRICES IN QUANTITIES

BOURBON RED TURKEYS.

The above is a splendid illustration of the Bourbon Red Turkey. You can see by it exactly what the bird looks like.

So many of my friends have written me asking about Bourbon Red Turkeys and the testimonials of those who have been raising them, lead me to start a flock last year; I like them just fine, it seems as if the Red color is a very popular one. We all know that the red and buff fowls hold a prominent place with the most critical fanciers, then there seems to be something about the "Red" that denotes hardiness, any one that has raised or knows anything about the "Red" Hog knows that it is one of the most hardy of the Swine family, all Red and buff fowls are hardy and easy to raise; this holds good with the Bourbon turkey as well, they are not only hardy and easy to raise, but they are docile, easy to keep at home, do not have that roaming disposition.

They are a real dark buff color and as far as shape and size are concerned resemble any other breed of turkeys.

If you are in the market for a turkey that is different from those that your neighbors and friends have and such that will not run off, get a start in this breed.

I have received so very many compliments on them from breeders that I just could not help but offer them to my friends.

I am sorry that I am not in a position this year to offer females, will have plenty next season; this, as above stated, is my first season and will need all the females for the foundation start. I have some nice Toms to dispose of, and will have plenty of eggs.

Price of Toms: $7.00 to $12.00, depending largely upon the weight.
Price of Eggs: $3.50 per 9; $6.00 per 18; $8.50 per 27.
Write for special prices in quantities.

WHITE ORPINGTON COCKEREL A FINE BIRD.

Dear Madam:— Rogersville, Mo., 12-16-'13.
I received the White Orpington cockerel all right today and think he is a fine bird. I am returning the coop today.

Yourstruly, J. N. BIGGER.

Kind Words
From a Few of Our Friends

KIND WORDS—THE MOST CONVINCING EVIDENCE IN THE WORLD.

This is an age of "Show Me," well I can do it all right and by the most convincing evidence is testimonials from disinterested customers. I ask you to read the testimonials that I show in this book. This is the kind of evidence that cannot be bought at any price. It is the evidence of **perfectly satisfied customers**, most of them who have never seen me and have no earthly interest in me beyond a friendly feeling and acquaintance brought on by the personal friendliness displayed in this book and my letters and because I have treated them on the **"Golden Rule Plan."** These people will not say they were satisfied and the stock, eggs, incubators, supplies or whatever they order were not all that I claim for them if it were not so, even for a friend; so could it be expected that they would do this for only a stranger? I have only room to show a very few of the very nice letters that I receive daily, but I am going to show a sample at least, and have not picked the most favorable ones, but show a true sample from my testimonial files which can be seen any day in my office.

And now dear friends, I want to express my heartfelt thanks for these kind words which have been such a source of great pleasure to me, I am sorry that I could not publish all the kind letters of hearty appreciation of the manner I carry out my version of the **"Golden Rule."** I wish to thank you one and all, both those whose letters I have published and those I had to pass up for lack of space, as I am only giving a few comparatively of the number I have received I wish that you would read every one of them, because it shows the good fellowship feeling that prevails between my customers and myself.

Little Winifred, daughter of Mr. and Mrs. Weiland. Now isn't she just as sweet as she can be? She reminds me of one of my little girls as she was a few years ago, but now in high school.

A PLEASED AND SATISFIED CUSTOMER.

Martinsville, Ill.

Dear Madam:—Enclosed find picture of my daughter, Winifred, feeding our 80 S. C. White Leghorns. I have 6 Partridge Plymouth Rock pullets; can you furnish me with a Partridge Rock cockerel? I think one that will be two years old next spring would be best for breeding purposes. Please advise me your plan, also prices, one male, also pullets.

Yours truly,
J. A. WEILAND.

FROM THE DENVER RECORD POULTRY EDITOR, DENVER, COLO.

(This is a clipping from the above named paper. It gives me rather a favorable write up, but shows how I am regarded by the press.)

When we see so many men and women spending their lives and energy trying to perfect the various breeds of poultry; when we see over half a hundred large Poultry Journals filled with advertisements of thoroughbred Poultry, does it mean that all these people have simply taken up the fad in order to occupy their time, or is it because they see that there is money in the business.

A Boston lawyer with a large practice has established and operates one of the largest poultry farms n the east. The great musician's wife, Mrs. Paderewski, has a great poultry ranch in France. May Irwin, who has delighted the best audiences on two continents with her fine acting, has a large poultry ranch on the Thousand Islands and says that she never enjoys herself half as well as when on her farm looking after her lordly pure bred flock.

While it may be a fad and an amusement for some wealthy people, it is safe to say that 90% of the poultry in the United States is raised by people because it is profitable; even the over-rich like to make money as well as other folks.

From the Agricultural Department of the United States we learn that the poultry crop is greater than any other farm products excepting wheat.

There are many large poultry farms, in the east and California, that clear from $10,000.00 to $25,000.00 annually. Poultry pays when handled on a large scale, and also when raised in the back yard.

BERRY'S GOLDEN RULE POULTRY FARM, CLARINDA, IOWA

Here is what one woman did last year:

In the fall she bought 100 hens, costing her $90.00. Feed for the year was $133.99; oil, $8.44; houses and coops, $26.86; total outlay, $259.29. During the year she gathered 10,608 eggs. With an incubator she hatched 3,267 chicks. Eggs and chickens sold for $529.37. She now has 120 pullets worth $120.00 and $9.40 worth of food on hand. Total income, $658.77. Deducting the expense, she realized a profit of $399.48. This means a profit of $4.43 on each hen; nearly $400.00 on a small flock. Here is a clear profit of over $1.00 a day for every day in the year. Many people work hard from daylight to dark for less money.

Women often succeed better with poultry than men. There are two lines in which women beat men all to pieces; one raising babies and the other raising chickens.

One of the most successful raisers of poultry in the United States is Mrs. A. A. Berry, manager of the Golden Rule Poultry Farm at Clarinda, Iowa. Mrs. Berry used to teach school, but wanting a broader field and more money, she turned her attention to poultry. Their farm covers 100 acres in Page County, Iowa, which for many years has raised more corn than any other county in the United States. Her improved farm land is worth from $200 to $500 per acre. It is probably the richest agricultural county in the United States. On this farm Mrs. Berry raises all kinds of poultry—chickens, turkeys, geese and ducks. She has also built up a large trade in incubators and poultry supplies.

One of the best known poultry judges in the United States, after visiting her farm, wrote that he was much surprised to find such a fine, splendidly equipped plant for handling pure bred poultry.

He pronounced her stock of birds as being the best that it had been his pleasure to examine. Mrs. Berry has also written several books on Poultry Culture. Her book designated "Profitable Poultry" is one of the best treatises it has been our pleasure to read.

Anyone desiring to start in the poultry business can get either birds or eggs for hatching as well as incubators and brooders from Mrs. Berry at reasonable prices and she takes great pains to give her customers such help and instruction as is sure to make their business profitable. If you want to start right in the poultry business with the best of foundation stock, it will certainly pay you to correspond with Mrs. Berry.

WHAT A GOOD JUDGE SAYS ABOUT OUR FARM.

Berry's Poultry Farm,
    Mrs. A. A. Berry, Mgr., Clarinda, Iowa.

Dear Madam:—I want to say to you that I appreciate my visit to your big poultry farm last week and I take this manner of thanking you for courtesies shown.

I was certainly very much surprised, agreeably so, at your fine plant and splendid arrangement for handling pure bred stock, eggs, incubators and supplies; but most of all was I surprised that you had such a fine stock of birds. You are certainly a splendid judge and a first class breeder to produce the kind of birds I saw on your place, and such a fine lot of youngsters just coming on; but they could not help being good, hatched from the fine stock in your breeding pens. I only wish that I could have stayed longer and visited your specialty plants that you have scattered around Clarinda.

I just wish to assure you that I will not fail to say a good word for the Berry's Golden Rule Poultry Farm whenever opportunity presents itself. I saw enough of your methods to know that you will please your customers and give them a square deal in every particular.

Wishing you the continued success your efforts deserve and again thanking you for your kindness, I remain,

    Yours very truly,
        C. A. SAYLER,
        Rep. Poultry Success.

Poultry Success Field Man and Noted Poultry Judge.

SO PLEASED AND SATISFIED THAT MR. ANDERSON SENT ME A FINE PHOTO.

Ellinwood, Kas., 2-23-'12.

Dear Madam:—

I received the cockerel in the best of condition and am thoroughly satisfied with him. Enclosed you will find photograph of some of my pure bred White Rocks and the cockerel received from you. This is a photograph taken by myself and I know that it is original. This is certainly a fine pen of White Rocks, and the cockerel is of such quality that I can expect even better birds next year than I have now.

    Yours truly,
        T. H. ANDERSON.

Do you see that prize White Rock cockerel and the head of a fine prize winning pen of birds? They belong to T. H. Anderson, Ellinwood, Kans., and he got the cockerel from me. He is a good judge and knows where to get good chickens.

Little Hume Hall and his "Berry Chickens" as he calls them. Read what his mamma has to say about the "Berry" strains of chickens. His mamma raised 85 of these white Leghorns from 100 eggs that she purchased of me.

RAISED 85 WHITE LEGHORNS FROM 100 EGGS. Proof that our Eggs Hatch.

Dear Mrs. Berry:— Thompson Station, Tenn.

I am sending you today two photos of part of the chickens hatched from eggs that I got of you this spring. The picture does not do them justice as they are so much prettier and nicer than the picture shows them. The little fellow in the picture is my oldest son Hume, 5 years old, he calls them the Berry Chickens, he says that he likes them best. We raised 85 of the single comb white leghorns out of the 100 eggs purchased and of the 25 buff orpingtons hatched 13 all living. I cannot say too much for your stock of chickens, they are so easy to raise. Yours truly, W. T. HALL.

GOLDEN WYANDOTTES MIGHTY FINE.

Dear Madam:— Eaton, Ind., Dec. 31.

I thought that I would drop you a line to tell you how much I think of the Golden Wyandottes, which came the 13th day of Dec. This was on Friday and on Friday, Dec. 27th, I got the first egg. I think these birds are mighty fine. A little later I want to get a Biddy Incubator and Brooder, also eggs. With many thanks for the fine birds you sent me, I am. Yours truly, JOHN A. KIME.

Above shows some of Mr. Kime's fine Golden Wyandottes, he and his wife are in the picture with them. Mr. Kime is just starting in the Golden Wyandottes and still has a few mixed chickens, he expects to cull them out. The mixed chickens also show in the picture.

WHITE LANGSHANS PROVE SHOW WINNERS.

A photo of Mr. Gardiner, Jr. and the White Langshans that took all the prizes. Are they not simply beautiful? The White Langshan is a winner all right wherever you put them.

McDonald, Pa., 2-26-'13.

Dear Madam:—I am writing now to let you known how the birds did in the last show. I had five of your White Langshans in the Canonsburg Poultry and Pet Show, and they were awarded four prizes. I got first cockerel, first, second, and third pullets. The cockerel that took first prize was the last one I got from you. I would like to know what you could let me have about 6 more White Langshan hens or pullets for, one year old hens preferred if you have them. Write and let me know as soon as possible, as I would like to get another pen mated up. I think you will be getting orders from some of my friends as I have recommended you to everyone I hear of who wants poultry. Hoping to have a good season this year due to the good stock I got of you, and hoping to hear from you in the near future, I remain.

Yours truly, MR. JOS. GARDINER, JR.

WAS OFFERED $25.00 FOR THE COCKEREL ALONE.

Dear Madam:— Fisherville, Ont., Canada, 3-18-'13.

Received the Partridge Wyandottes today, and they are the most beautiful birds I ever saw. I never expected anything like I got. One man offered me $25.00 for the rooster alone. But I did not accept, as I thought I could not get another as good.

You have treated me so honestly and fairly, that you will receive an order from me next fall. Yours truly, RUEBEN NABLO.

WELL SATISFIED WITH THE PAIR OF CORNISH INDIAN GAMES.

Dear Madam:— Swissvale, Pa., 3-18-'13.

We received the pair of Indian Game fowls, and they arrived in good condition. We are very well pleased and satisfied with them. I am shipping the crate back as per your instructions. Yours very truly, JAS. MILDON.

MUCH PLEASED WITH THE BUFF COCHIN BANTUMS.

Dear Friend:— Logan Utah, 3-14-'13.

Received the trio of Buff Cochin Bantums and am much pleased with them. Will perhaps need some eggs later on. Thanking you kindly for your prompt attention, I am.

Yours truly, C. T. THOMAS.

ROSE COMB WHITE LEGHORN COCKEREL IS A BEAUTY.

Dear Madam:— Hale, Mo., 3-3-'13.

I received the R. C. White Leghorn cockerel in fine shape and he is certainly a beauty. I will return the crate in a few days. Yours truly, I. M. THOMAS.

A view of Mr. Gardiner's White Langshan pens and some of his fine birds. I think they are simply beautiful, don't you? Too much cannot be said in favor of the White Langshans, they are the coming fowl.

THE PARTRIDGE COCHINS THE MOST BEAUTIFUL BIRDS EVER SEEN.

West Point, Nebr.

Dear Madam:—We have received our Partridge Cochins all O. K. and I was so pleased when I saw the chickens. They are the most beautiful chickens I have ever seen. Our neighbors think they are very pretty, and I suppose they will send for some too, later on. I will send the coop back as soon as I can. Hoping to see the hen arrive safely, I remain, Yours truly,
JOHN SILA.

HOW IS THIS FOR A RECORD WITH THE "BIDDY" BY AN INEXPERIENCED CUSTOMER?

Donald, Wis., 5-14-'13.

Dear Mrs. Berry:—

I am enclosing a photo of my "Biddy" recently purchased of you, and my first hatch. I got 80 fine strong chicks from 102 eggs. Not a cripple or weak one in the bunch. And I am taking off my second hatch today, which is turning out still better. I have 81 good lively chicks from 92 eggs, and have 3 eggs left in the incubator which have pipped, and which I think will hatch and live.

You certainly make a fine incubator, when it will do work like this for a person who has never used any kind of an incubator before. I shall use the "Biddy" incubators entirely.

Very truly yours,
R. E. PROSSER.

RUNNER DRAKE AND HAMBURG COCKEREL SATISFACTORY.

Iron Hub, Minn., 2-10-'13.

Dear Mrs. Berry:—

We received the drake and the S. S. Hamburg cockerel all right and they are very nice birds and am well pleased with them. Will you send me your book on ducks as I never raised ducks, and want to know all I can about them. Yours truly, MAUDE TRAYS.

Mr. Sila, and the pair of Partridge Cochins he purchased of me. Are they not beautiful? A fine looking young man, too.

Mr. Prosser's "Biddy." Read what he says about the Incubators purchased from the Berry Poultry Farm. He has such a sweet, dear little girl.

Please study this photo. Some who have the desire do not enter into the poultry business because they think they have not enough space. This picture was sent by a customer who knew how to utilize the small space at his disposal. You can do the same. Much money can be made where some consider the area too small.

One of our good customers sent us this photo, but we fail to find her name. These are a few words she wrote on the photo she sent us and failed to sign her name. "My little nephew of La Fayette, Ind., is visiting me and I had this picture taken with his little friend feeding my single comb brown leghorns, the first thing he said when he saw the chicks was, 'Oh, Aunt, where did you get so many little beauties?'"

SO WELL PLEASED, EXPECTS TO SEND ANOTHER ORDER.

Atlantic, Ia., 1-30-'13.

Dear Madam:—

My order of one S. C. Rhode Island Red cockerel arrived safely, and in good condition. I am certainly well pleased with the bird. Have returned the coop as you requested. Hoping to send you another order soon, I am.

Yours truly,
B. H. WILLIAMS.

WELL SATISFIED WITH BERRY'S CHICKENS.

Platte Center, Nebr., 8-7-'13.

Dear Madam:—

Please find enclosed a picture of our little Zelma with her pet chicken. Zelma is two years and four months old, and is very proud of the chickens that I got for her, this spring, from your farm. We are well satisfied also. With best wishes for your success, I remain.

Yours truly,
A. J. GLADOWSKI.

SATISFIED WITH THE COCKEREL.

Waukesha, Wis., 1-25-'13.

Dear Madam:—

Your's at hand, and received the cockerel on the 24th in good shape and all O. K. I am well satisfied with him. Hoping some time to give you another order, I am,

Respectfully yours,
JAS. B. REED.

GEESE ARE FINE.

Butterfield, Minn., 1-28-'13.

Dear Mrs. Berry:—

The geese were received the 27th. They were all O. K. They certainly were fine birds. Thanking you for your shipment, and with best wishes, I am,

Yours truly,
MR. PETE ANDERSON.

WHITE LANGSHANS SATISFACTORY.

Peckville, Pa., 1-31-'13.

Dear Mrs. Berry:—

We received the five White Langshans today and are well pleased with them. We got one egg the day they arrived. When I order from you again, I want you to send me a setting of your best Buff Cochin eggs.

Your friend, MR. LOUIS RICHARDS.

Little Zelma, with her pet, her mamma got for her from Berry's Farm.

THE BEST BUFF ORPINGTONS HE HAD EVER SEEN.

Greenville, Mich., 12-29-'12.

Dear Mrs. Berry:—

I am sending you a picture of my Buff Orpington cockerel and the pullet that laid the first egg. I have received one dozen eggs from the pen since they have commenced to lay. I weighed my cockerel today and he tipped the scales at 8 lbs. A big poultry man was down to see them Friday. He says they are the best Buff Orpingtons he has ever seen.

Yours truly,
L. V. PICKELL.

ROOSTER IS BEAUTY.

Hegewisch, Ill., 2-4-'13.

Dear Friend:—Received the rooster in good order, and am very much pleased with him. He is certainly a beauty. I will send you the money for brooder and eggs later on in the month. Thanking you I am.

Yours truly,
MRS. CHAS. LARSON.

HENS COMMENCE LAYING AT ONCE.

Ijamsville, Ind., 4-7-'13.

Dear Madam:—Coop of chickens arrived this A. M., all O. K. and am sending the coop back this evening. Chickens are fine and well; two of the hens laid today after removing from coop and placing in poultry house.

Thanking you for past favors, I remain.

Yours truly,
LOGAN H. STEWART.

This is Mr. Pickell with a pair of Buff Orpingtons he received of me. Read what he has to say about them.

READ THE FOLLOWING CLIPPING FROM A LOCAL PAPER AT CONWAY SPRINGS, KAS.

Mrs. Julia A. Little entered Indian Runner ducks in the Poultry Show. There were five entries in this class and Mrs. Little won first premium on cock, first on hen, and, first second and third on pullet. These ducks are from the Berry strain of Clarinda, Iowa, and Mrs. Little's success with this strain of ducks is exceeding her most sanguine expectations. She shipped eggs to many points in Kansas and other states last year and is preparing to supply a greater demand this year.

This is Mrs. Thomas Bullock of Hocking, Iowa, with her magnificent pen of Silver Laced Wyandottes. This pen certainly shows quality. We sent this pen to Mrs. Bullock and she was highly elated with our selection. Don't you think you would have been?

PLEASED WITH THE SELECTION MADE.

Blairsville, Pa., 4-28-'13.

Dear Mrs. Berry:—I received the chickens all O. K. and am very much pleased with them. There have been several people looking at them and they all think them fine birds. I shipped both coops back, the 26th. I am glad to tell you that the pullet that has been sick is getting in good health again and I expect her to begin laying very soon. The other hen has layed 24 eggs in 28 days and I think that is as good as any of them can do. Thanking you very much for the selection of birds you sent me, I am.

Yours truly,
C. S. LONG.

BIRDS ARE CERTAINLY GRAND.

Gowanda, N. Y., 5-2-'13.

Dear Friend: — Reeived the birds all O. K. and am well pleased with them. They are certainly grand.

Yours truly,
MR. CHAS. J. HUFF.

J. N. CONGER, WYOMING, ILL., THE MAN WITH AN INDIAN RUNNER DUCK THAT LAID 4 EGGS IN ONE DAY.

Wyoming, Ill., 6-2-'11.

Dear Madam:—Your letter of May 31st just received, and in reply I wish to say that the duck that laid four eggs in one day was hatched from eggs that I bought from you last season. I have a number of young ones from her eggs and think very highly of them.

Yours truly, J. N. CONGER.

WHITE LANGSHAN COCKEREL A FINE BIRD.

Cassville, Mo., 4-4-'13.

Dear Madam:—I received my White Langshan cockerel the 8th, and was well pleased with him. He is a very nice bird. I returned the coop the same day.

Yours truly, MRS. NETTIE LAUDERDALE.

"PROMPT ATTENTION TO ORDERS" OUR MOTTO.

Scranton, Ia., 4-2-'13.

Dear Madam:—Received the eggs and incubator all O. K. Thanks for being so prompt. Yours truly,
JAMES C. WRIGHT.

The picture at the right shows a beautiful pen of fawn and white Runner ducks owned by M. P. Hayes, Dallas, Texas. He has the Berry strain, and they are fine.

HAMBURG COCKEREL, O. K.

Redding, Cal., 2-16-'13.

Dear Friend:—I received the fine Silver Spangled Hamburg cockerel all O. K. today. I am well pleased with him, and I think he is a fine bird. I returned the crate today. Thanking you very much for your prompt shipment, I am.

Yours truly, THOS. W. McLAUGHLIN.

The photo at the left is of Louis Rosa of far away Honolulu, Hawaii, taking off a satisfactory hatch from the "Biddy." Mr. Rosa has bought of us "Berry's Biddy" incubator and several shipments of chickens, consisting of several different breeds. He received the last shipment the fore part of October and writes us that he is more than pleased with our selection. He is making ood.

BERRY'S GOLDEN RULE POULTRY FARM, CLARINDA, IOWA

Mr. Brown and his ducks that commenced laying at four and one-half months of age. Just read what he says about the Berry strain of Runners. Wonderful, but true.

DUCKS COMMENCED TO LAY AT 4½ MONTHS OF AGE.

Dear Madam:— Clarksboro, N. J., 10-6-'13.

Enclosed you will find a photo of a bunch of my Runners. They are four and a half months old, and have commenced laying. Some of my friends tell me that ducks will not lay in the winter, they think it is out of reason for them to lay at this time of the year. The old ducks stopped laying in August, but they are at it again laying most every day. And to say I am proud of my ducks, is putting it very mild for they are beauties. Give me the price on one Insurance fireless brooder.

Yours truly, R. S. BROWN.

GANDERS SEEM TO BE ALL RIGHT.

Dear Madam:— Border, Wyo., 2-3-'13.

I reeived the two ganders all O. K. and they seem to be all right. I will return the crate some time this week as I will be going to the express office this week some time.

Hoping they will prove to be good breeders, I am,

Yours truly, S. W. CONDRON.

Mrs. G. W. Harris, Shelby, Ohio, and her $500.00 Buff Orpington hen. That's its value. Read what she says.

OUR CUSTOMERS RAISE $500.00 BIRDS. MR. AND MRS. HARRIS RAISE VALUABLE BIRDS.

Shelby Ohio.

Kind Friend:—I herewith enclose a picture of myself and a hen that Mr. Harris values at $500.00. He thinks that she would score 97. We are certainly well pleased with the chickens received from you, they are splendid layers. Your strains of fowls are fine in every particular.

Yours truly, G. W. HARRIS.

RECEIVE THANKS FOR SENDING SUCH A NICE BIRD.

Chinook, Mont., 2-4-'13.

Dear Madam:—Received the cockerel all O. K. and many thanks for such a fine bird. Have sent the coop to Chinook to be returned to you.

Respectfully, MINNIE WALKER.

EGGS PACKED WELL.

Stronghurst, Ill., 4-22-'13.

Dear Mrs. Berry:—The eggs came thru in fine shape. The way they were packed, nothing short of a collision could have broken them.

Yours truly, MRS. ELLEN FINCH.

BUFF ORPINGTON COCKEREL SEEMS TO BE RIGHT AT HOME.

Boulder, Mont., 2-14-'13.

Dear Madam:—The Buff Orpington cockerel received in good condition and am well pleased with the bird. He seems to be right at home. Am returning the crate today. Thanking you for your promptness, I remain,

Yours truly, THOS. INGLING.

Is not this a beautiful flock? Mrs. Hopkins of New York State raised them from only a trio, one male and 2 females. You can do as well. All you need is a start of the right kind.

A SPLENDID RECORD OF A TRIO OF RUNNER DUCKS.

Blasdell, N. Y., 6-21-13.

Dear Mrs. Berry:—

We enclose you herewith photo showing part of the ducks we raised from the trio we received of you. We succeeded in raising 125 ducks altogether. We had splendid success with them, they are very strong, healthy ducks, and we only lost three, and those when right small.

Yours truly,
MRS. P. HOPKINS.

PLEASED WITH BARRED PLYMOUTH ROCKS.

Stella, Nebr. 12-6-'12.

Dear Friend:—The Barred Plymouth Rocks arrived in good condition and am well pleased with them. Will return the coop by express today as requested.

Yours respectfully,
HENRY SAYER.

WHITE ORPINGTON CERTAINLY A FINE BIRD.

Alexandria, Mo., 2-23-'13.

Dear Madam:—

I received the White Orpington cockerel this morning. He is certainly a fine bird. Many thanks for your pointers in selecting the pen. I returned your crate today.

Yours truly,
S. D. SEYMOUR.

Below is a pretty little picure sent to us by one of our good customers whose name and address we fail to find. In the few words that we find on the photo she says that her yard is so shady that it is hard to get a good picture. She also states that in the picture are the Anconas, Hamburgs, white and brown Leghorns that she purchased of me. You see that nice looking dog by her side, she says that is "Fritzie" and he goes with her to the pens every time she goes and if one of the birds gets out Fritzie puts it back in. That is surely a nice kind of dog to have.

Progressive people always look a little ways ahead, I wonder how many good people who look into the little fellows face pictured above, see the same thing that I do. I see one of the most promising and prominent poultrymen of the coming generation. His mamma, Mrs. J. A. Berry of Glen Allen, Mo., thought that he looked so much like my boy Earnest, whose picture I had taken by the pumpkin, that she just thought she would send it. They do look a great deal alike and the strange coincidence is, he is a "Berry" too.

A photo of Mrs. Casey's flock, the offspring of two ducks. Raised 75 from two ducks in one season. Is it not convincing that Runner Duck Culture is profitable?

READ THIS WONDERFUL RECORD. DOES IT NOT SEEM MARVELOUS WHAT TWO DUCKS CAN DO?

Dear Mrs. Berry:—   Correctionville, Ia., 1-12-'13.

Will write you a few lines for your next year's book. The trio I got of you surely proved very satisfactory to me. They laid just fine and the eggs are the best I ever ate, and the ducks are very delicious. I raised 75 from the trio, and hatched lots more but the Spring was so wet and cold that I lost a good many. I don't think I shall hatch so early another Spring. Will say they are all you recommend them to be, you cannot praise them too highly.

Your friend,
MRS. C. G. CASEY.

TURKEY A FINE BIRD.

Moscow Mills, Ia., 4-12-'13.

Kind Friend:—Received the tom in good condition and think he is a fine bird. Am well pleased with him. I thank you very much for your trouble. The coop was shipped back to you today. Yours truly,
MRS. E. A. POLLARD.

THE BIDDY IS A FINE MACHINE.

Spring Arbor, Mich., 4-12-'13.

Dear Madam:—I received your incubator April 3rd, and can truthfully say that it is a fine machine. To try it out, I set it with some of my own eggs, Buff Orpingtons. I will send you an order for some eggs after this setting is hatched. I have had my picture taken but it is not developed. People who have seen the Biddy like it very much. Hoping to hear from you soon, I remain.

Yours truly,
Deane Chambers.

EGGS PACKED GOOD AND RECEIVED EXTRA ONES.

Manchester, N. H., 4-9-'13.

Dear Madam:—Hatching eggs received April 5th in good shape and must say that you take great pains in packing them. Thanks for the extra ones. Hope they will hatch as well as they were packed. Will write you again after the hatch comes off.

Yours respectfully,
D. A. ROGERS.

This is Mr. Samuel McKinley, of Burnham, Pa., who hatches lots of chickens in a "Biddy" and has a fine flock of White Leghorns. He has been a good customer for years and is always ready to say a good word for Berry's Golden Rule farm.

A PROMISING START FOR ANOTHER YEAR.

Dear Mrs. Berry:— Rudd, Ia.

We enclose a picture of Rupert and Reuben, our twin boys, and their 8 Runner ducks. The ducks are just as nice as they can be. We expect to get a good start from this pen another year. Yours respectfully,

MRS. W. R. THATCHER.

ANOTHER SATISFIED CUSTOMER.

Dear Madam:— Valeda, Kas., 5-3-'13.

I just received my Hamburgs and Wyandottes last night, and am very well pleased with them. The Partridge Wyandottes are sure beauties and the other varieties are very nice birds. I am well pleased all around with the shipment, and thank you for filling the order so quickly. I sent the coops back this morning. Hoping that I may write you good news later regarding my birds, I remain.

Yours truly,
ROY E. JAMES.

PARTRIDGE WYANDOTTES RECEIVED O. K. AND APPRECIATE PROMPT SHIPMENT.

Dear Madam:— Sioux Center, Ia., 2-3-'13.

Received the pen of Partridge Wyandottes this morning, and am very much pleased with them. I also wish to thank you for sending the order so quickly.

Yours truly, JOHN B. DOORNWAARD.

Rupert and Reuben Thatcher, the twins, and their ducks. Fine boys, all right.

DUCKS DOING WELL.

Dear Madam:— St. Clair, Minn., 12-6-'12.

Your shipment of ducks to me arrived all O. K. and they are doing well.

Yours truly, J. SHIELDS.

HOUDANS WIN FIRST PREMIUM.

Dear Mrs. Berry:— Hopkinsville, Ky., 10-10-'13.

I won first premium on a pen of Houdans at the Pennyroyal Fair yesterday.

Very respectfully, MRS. MINNIE BRASHER RENSHAN.

AN EXTRAORDINAY GOOD HATCH ESPECIALLY WHERE THERE ARE SEVERAL VARIETIES HATCHED TOGETHER UNDER THE SAME HEAT.

Dear Mrs. Berry:— Columbus, Ohio, 4-12-'13.

Received the eggs in very good condition, considering the distance. There was one broken on top, but the rest were all right. We hatched 5 Anconas, 4 Hamburgs, 4 Brown Leghorns, and 4 White Leghorns. The eggs were all fertile, but two. I hatched the eggs under hens and raised the chicks in a brooder. Have not lost one yet. Have 45.

Thanking you for your promptness and kindness, and hoping to hear from you again some time, I am. Yours very truly, MRS. U. M. STEINBARGER.

Mrs. Steinbarger and her husband with some of the chicks hatched from eggs she received from me. Fine people all right.

About Incubators

WILL IT PAY TO BUY AN INCUBATOR AND BROODER?

I think that this question is so generally understood that I do not thnk it necessary to devote much time and space on this subject, but for the benefit of the "doubting one" will give some reasons why it will pay you to buy an incubator and brooder.

While some might doubt if it will pay them to buy an incubator and brooder, I think that every one knows that there is no question about the great value and advantage of a good incubator and brooder. Just as well ask the question will it pay to buy any of the numerous labor saving devices that are in daily use.

Modern machinery, improved implements and household devices have enabled one man or woman to do the work of many, and there are thousands of labor saving and money making inventions. Just so with incubators and brooders; they have enabled the poultry raiser to lessen the labor, the risk, and do much more of it than by the old hen method.

The greatest profit (and money is the principal thing) we are all after in the poultry business, and you can accomplish it better with a good incubator and brooder, instead of using the hens. It will take 25 or 35 hens to do the work of a 360 "BIDDY" incubator. You can have chicks when you want them, as you do not have to wait until the hens take a notion to set. This enables you to get earlier chicks and a better price for them and your pullets will make better winter layers.

A hen is uncertain even when she takes a notion to set; as she may desert the nest after setting a week or so. The "BIDDY" is always ready for business and sticks to its job. A hen may be clumsy or nervous and break the eggs. Our "BIDDY" does not.

Hens cause much anxiety because one is always afraid she will leave the nest, break the eggs or eat them, and is much more trying on the nerves than an incubator, as you can see at a glance how the temperature is running and can see that everything is O. K. and will remain so, and hatch all fertile eggs.

One incubator can be managed more easily than ten hens, and with the "BIDDY" incubator a higher percentage of fertile eggs will hatch. You can hatch all kinds of eggs, such as turkey, geese, ducks, and any other fowl with the "BIDDY."

It will pay for itself twice over each year and will last a lifetime, but the best argument I can make is to state the fact that thousands are using incubators and brooders and make money doing so and like them. From reports by different factories there were over 350,000 incubators sold last year, which was an increase over the previous year; in fact, the number increases very much each year. Just think of this great number, if they were not a good thing and paid well, would this be true?

You have to get a good one to compete with these thousands of incubator users.

It pays others, it will pay you.

GREAT ADVANTAGE OF AN INCUBATOR.

We will here name some of the good reasons why you should buy an incubator:

Because it will pay you to do so, as you can raise earlier chickens and more of them and at less cost than by the old methods. Now you will say, prove that statement.

The greatest argument I can advance is that thousands and thousands were used by people that know they are a good thing and it pays them. If it did not pay them they would not use them year after year.

The incubator is becoming more popular and sales are increasing very rapidly, but the day of cheap-skate machines is past. People want the best; they want a good incubator and want it as cheap as is possible to obtain a first class machine.

There are a good many machines placed on the market that are a detriment to the incubator business. They make a person dissatisfied, as they cannot use them successfully, so tell their neighbors that incubators are all a failure.

They condemn them all because theirs was a failure. Don't let any one tell you that incubators are a failure; because as a class they are not, but a fully established success.

If there is no successful incubator in your neighborhood, or if you want to make the chicken business more profitable, get a "BIDDY" and show them.

There is no question that you can do it and make a success of it.

HEATING AN INCUBATOR.

This is the essential thing about a hatchine machine, to produce the same regular heat that a hen does, when she is setting. Uniform heating is the whole thing about an incubator.

Some claim that hot air forced through pipes is the best; but they have little argument to back their claims.

A machine like that can be made cheaper, as they can use tin or galvanized iron pipes, there being no water to rust them out; such material costs four or five cents per pound, while copper costs 25 to 40 cents per pound.

Hot water forced through copper tanks and pipes will keep an even temperature, easier to regulate and cheaper to warm, and will keep the proper temperature, and will produce a moistening temperature such as a setting hen produces. If your light for some cause has gone out the heat retained in the water and pipes will keep up the temperature quite a while after the lamp goes out. Not so with the hot air machine. Just as soon as the lamp is out the temperature begins to fall, as there it nothing to hold the heat.

Every one knows that the best heat for a building is hot water or steam, although it costs more than hot air; but there is no question about it being better. Quite frequently they force steam and hot water long distances through pipes. A good portion of some cities is heated from one central plant. In our little city of Clarinda they have a heating plant where a good portion of the business part and residences for several blocks are heated by forcing steam through pipes and radiators established in the buildings, which heats them much cheaper and better than they could in any other way. If they would try to force hot air, it would have to be so hot at the plant it would blister the pipes and then could not get it warm enough to carry a block to do any good towards heating a building, as the far end would not be supplied with sufficient heat.

Just so with an incubator; a hot water machine has a heating tank and forces hot water through pipes in the egg chamber and an even heat is supplied throughout.

Don't let these hot air people fool you into thinking theirs is the best, because it is not common sense, and experience has taught us that there is only one kind of heat that is practical and economical. We use hot water in the "BIDDY" because we know it is the best, although it costs much more.

HOW TO CHOOSE AN INCUBATOR.

Now I know that this is sometimes a very serious question with so many to choose from, and each one says theirs is the best.

I know there are good incubators besides the "BIDDY"; but I can prove to you that it has some advantages and superior points over all others. In choosing an incubator: First consider the reliability and responsibility of a firm and then read the description carefully of the machine they are offering. Do not let gush and jolly talk influence you; actual facts and sound arguments from a reliable source backed by good testimonials and recommendations from reliable parties or firms are worth a whole volume of unreliable statements.

In choosing an incubator consider the ability of the parties offering it. Consider the practical experience they have had in running and handling an incubator and the actual knowledge of the poultry business from the ground up. Above all, consider the simplicity of the machine, as the simpler and fewer parts a machine has, the easier it is to operate, so will give better satisfaction.

Be governed by your own good sense and business judgment. After investigating thoroughly I am willing to have you put the "BIDDY" to the closest test and I know the verdict will be favorable.

WHAT SIZE IS BEST?

There are a great many questions asked me in regard to the best size to purchase, and the capacity that I would recommend.

If I knew your conditions and circumstances, I perhaps could be called upon to answer your question; perhaps in a general way, I might give you my ideas and then you can work from that to the best of your ability.

If you will stop to do a little figuring, you will be convinced right on the start that the largest machine is the cheapest machine, because it holds more eggs according to the price than any other machine, and the 360 machine is a splendid machine for those wishing to go into the poultry business exclusively, or wish to make it a good side line.

The 240 egg capacity is a splendid size, and will say that we sell more of that capacity than any other. The 180 capacity is also very popular and the best size for a great many conditions. The smallest or 120 capacity is a splendid size for any one going into the business on a small scale, or for the man with a city lot. We would not advise bothering with a smaller machine, because it would cost almost as much as the 120, and what would be the use bothering with an incubator that would not do the work of more than 4 or 5 hens.

WOULD NOT GIVE THE BIDDY FOR THREE OTHER INCUBATORS.

Dear Madam:— Elmira, N. Y.
I put 220 eggs in the "Biddy" Incubator and had 161 fertile eggs, got 135 healthy chicks. If the other 26 had been strong in fertility like the 135, I would have had 161 chicks.
I will gamble on the "Biddy" hatching every strong, fertile egg. I have three other makes and I would not give the "Biddy" for all three Incubators. The capacity of my other three machines is 765 eggs. Yours respectfully, W. R. ADAMS.

CHEAPNESS OR QUALITY?

This is a big question with many and it is a serious question, this matter of price.

We all like a bargain and to get things just as cheap as possible. They poke a good deal of fun in the newspapers about women as bargain hunters. Perhaps they are closer buyers than men, anyhow they are not going to pay more for a thing than they have to, and nothing pleases them more than a good bargain.

I think that most of the men are that way too; but I do not think that women would buy as many worthless things because they are cheap or take a notion to them as the men will.

Any one knows that flimsy made racket store goods are dearer in the long run than articles made well and of material that cost a little more. Every one knows that there are many things made nowadays in the battle of fierce competition with a view to cheapness and low price that are utterly worthless. There are lots of machinery that are made so poor and of such worthless material that are not worth taking home or fooling with.

I am sorry to say that there are some incubators and brooders made that are a disgrace and a detriment to the business. Poor wood, unseasoned, which warps and cracks; tin, galvanized, or very thin copper tanks; poor regulators, and a machine that will not give good satisfaction or last long; but this should not influence any one that incubators are not a good thing any more than an $8.00 sewing machine or a $30.00 top buggy should influence any one against these things. Any one knows that good articles of these things cannot be made at those prices, and yet they are made and lots of them sold, but they are made to sell and will not last long or give satisfaction.

I could cite you an incubator advertised for $3.98. How is it made? Cottonwood or cheap pine box, single wall not any better than a soap box, tin tank, cheap lamp and a machine that would not last a season or hatch one out of every five fertile eggs. It is a machine that is dear at any price, and made to sell.

In buying an incubator, GET QUALITY, not cheapness; get a good incubator and get it as cheap as you can, but be sure that the quality is there, as there is nothing so unsatisfactory as a poor incubator and brooder.

I am willing that you should judge the "BIDDY" by the closest test, and you will come to the conclusion that the "BIDDY" is the best and the price is right and surely is a bargain.

HOT AIR OR FACTS; WHICH?

It is really disgusting, the amount of hot air and wild statements some of these incubator catalogues give out, and in many cases it is done so cleverly and smoothly that it often fools the buying public.

It is often hard to distinguish between these wild and extravagant statements and cold facts.

I know that every one wants the facts about the articles they are buying, wants reliable, honest statements and truthful description that they can rely upon. The great detriment of the mail order business today is that the law of our land does not punish the exaggerator, the magnifier, and the misrepresenter. It is hard to determine between this class of statements made in their flowery catalogue and the reliable truth that can be relied upon.

I know that you want the cold facts and truthful description. I know that you want a square deal.

I know that if you do not get a square deal from me that I cannot expect to get any more business from you, and as I am in the business to stay and want to sell you poultry and eggs or supplies again, just as often as you want them.

The incubator firm who only makes incubators and brooders cannot expect to sell but one to the most of the people, and some of them, I am sorry to say, are not particular as to the quality of the machine they send out, knowing well that but few buy more than one. Not so with me. I want to sell you poultry, eggs and supplies, just as often as you want them.

I have built up this business from the bottom and made a success of it and you know that this only can be done by doing business on the Golden Rule plan, a square deal to all, and I expect to continue to do business in this way—to give big value for your money and make every one a pleased customer.

Knowing these facts and pursuing this plan, I cannot afford to give anything but plain facts in my descriptions of what I offer.

I give lots of good references in regard to our reliability and ask you to read the letters from our pleased customers as to my treatment.

I want your confidence and expect to keep it by honest, square treatment. You are to be the judge of whether I am making wild, exaggerated statements or facts. I want to know you and do business with you this year, and if I get a chance to I know that I will again.

Berry Biddy Incubator and Brooder

A truly successful person is progressing and making progress each year. I am successful and progressing and improving in every way each year, and I know we have a machine in the "BIDDY" that is ahead of anything I have tried or know of and I think that I have tried or know of the most of them.

I have added a few new features to our "Biddy Incubator" this year that I know are good and a great improvement. I have perfected the incubator so that it is so very sensitive that one breathing upon the thermostat the damper will be affected. The regulator is the mainspring of any machine. We have installed a larger and deeper nursery which is a decided improvement. We also have brought into use a deeper egg tray, so that the eggs can be turned and handled without danger of breaking or knocking them out of the tray. Notice the pictures of our machines very carefully. The illustration showing the 240 egg machine, shows exactly how all of our machines are built.

The "BIDDY" is made of the very best material that is known, and experience has taught that California red wood is the best wood, that heavy copper is the best for tanks and heater, and I use the very best grade of each; in fact, everything is the best, the workmanship is of the best and most skillful that money will obtain. Nothing is omitted in the way of making the "BIDDY" the very best.

The design and improvement are superior to any other machine made and makes it the most durable, easiest to operate and will hatch more chickens. The brooder raises more chicks and lasts longer. We have spared neither time nor expense in making the "BIDDY" the highest grade machine in every way, and are offering it at a price that is lower in comparison with any other.

You will make no mistake in buying the "BIDDY."

OUR 1914 MODEL "BIDDY."

I do not hesitate in saying that our 1914 "BIDDY" is the best in every way and the price is right and a bargain.

I am using the best material known in its construction. The most skillful workmen only are used. New and up-to-date machinery enables us to sell you a better machine for less money than any other firm. Tight walls, plenty of space for eggs and nursery below, scientifically constructed to furnish just the proper amount of moisture, simple, but very effectual, end regulator that regulates itself, large tank and heater that distributes just the right amount of heat, best thermometer made and we candidly make the claim that it comes very near perfection in an incubator. But perhaps you will say, "so and so makes a big claim for theirs," and that they all say "theirs is the best." Well, I think that I prove that the "BIDDY" is **the best,** or at least equal to the best, and costs less money. Read the descriptions and arguments as to being the best. Also note my iron clad guarantee.

SIMPLICITY AND EASE OF OPERATION.

We make this a strong feature in the "BIDDY." You do not have to sit up nights to keep an even heat. The machine is so constructed and the regulator is so easy to adjust that after you get it started it does not have to be changed; but

SEE HOW MR. McKINLEY IS GETTING ALONG WITH THE "BIDDY."

Burnham, Pa.

Dear Madam:—I will drop you a few lines and let you know how I am getting along with the "Biddy." It is certainly working fine. The first hatch produced 80 per cent, which I call extra good. The next batch is pretty well on the way, and the prospects are excellent. I want to get a pair of Rose Comb Brown Leghorns; please give me your best price.

Yours truly,
SAMUEL McKINLEY.

will keep up an even temperature. Just fill the lamp and turn the eggs as per printed instruction with each machine and it will do its own regulating and supply a sufficient amount of moisture to hatch all fertile eggs.

Some of the best hatches are made by inexperienced people, the first trial. It is so easy to operate that quite frequently an operator becomes careless after a few hatches and forgets to fill the lamp and it goes out, or forgets to put the eggs back in the machine when cooling them, until the next time to turn them. Of course, this is gross carelessness. There is no question, but it is a fact that any one can run the "BIDDY" and they don't have to stay home from church, social or other meetings to look after it, or sit up at night to keep it regulated.

THE "BIDDY" IS SOLD ON ITS MERITS.

My iron clad guarantee protects you and is surely proof that the "BIDDY" is sold strictly on its merits. It must be meritorious and do what I say it will or you do not have to keep it.

Our experience on the Golden Rule Poultry Farm has proved to us that it has merit, that it has quality, that it will do just what we say it will, as we know it has for us, therefore you take no risk in sending for a "BIDDY."

Mr. and Mrs. J. H. Fenton, Havre, Mont., and their pretty little girl who is helping to take off the biggest hatch ever taken off in Montana from 152 untested eggs. Now isn't that a pretty sight?

A SPLENDID HATCH IN MONTANA—SEE THE PICTURE.

Dear Madam:— Havre, Mont., 6-9-'11.

I received yours of the 30th ult., and also the Indian Runner Duck eggs, only one was broken. I am enclosing you herewith a postal photo of the "Biddy's" last hatch. We hatched 136 out of 152 eggs set, and every chick is alive and strong so far, not losing even one. We did not test out the eggs. Thanking you for past courtesies, I remain,

Sincerely yours, MRS. ERMA FENTON.

THE "BIDDY" IS BACKED BY EXPERIENCE.

We use it on our mammoth poultry farm. It hatches more chickens than any other kind we have tried. We have improved it to its highest state of perfection by operating and making numerous tests. It is practical, not a theory. Experience has been our teacher and you know that this is the best in the world, although the most costly.

Everyone that buys a "BIDDY" gets the benefit of our years of experience in hatching and raising of chickens.

THE "BIDDY" BACKED BY THE STRONGEST GUARANTEE.

We make just as strong a guarantee as can be made. We have absolute faith in the "BIDDY," we know just what it will do for any one that knows how to run a washing machine, sewing machine or churn. It is not nearly so complicated as a sewing machine or as difficult to operate.

We know what it does for others and know what it will do for you. So I do not have any fears in backing the "BIDDY" with a strong guarantee.

IRON CLAD GUARANTEE.

I guarantee the "BIDDY" incubator and brooder to be fully as represented. You may use the "BIDDY" three hatches and I guarantee it to give good results if it is given a fair trial and the directions followed. If it does not fill the requirements, notify me and I will remedy the defect, replace without any cost to you, a new machine or refund the money. I also warrant the machine for five years, if proper care is given it. MRS. A. A. BERRY.

No strings to this; it must do as we say it will or you come back to us. You do not take any risk. I take the risk, it has to work.

Now, what can be fairer or how could we make any stronger guarantee? Surely you would not hesitate in buying the "BIDDY" for fear that it will not work or that you cannot operate an incubator? We take that chance as we know the machine will work and that every one with common intelligence can operate it successfully. So we guarantee it for five years, that it will not warp or crack or the case open in any way, if proper care and attention is given. The heater or tank will not rust or wear out. It is built so substantial that it simply cannot go wrong, barring accidents. I do not think you can wear it out in a lifetime, if you give it a fair treatment, unless it is that some minor matters, such as burner or wicks, which cost but little; but these will last a long time if proper care is given them.

WE PAY THE FREIGHT.

We deliver the "Biddy" at your railroad station so you do not have to worry over excessive freight charges. We make you a delivered price so you know exactly what it costs you. Our railway facilities are excellent for quick service, although we strongly advise you to order in time, if possible. When you get a "BIDDY" incubator and brooder you know just what it is going to cost you; this is a great advantage that should count for much in choosing the "BIDDY." We pay the freight only on incubators and brooders.

THE "BIDDY" THE BEST FROM EVERY POINT OF VIEW.

It combines quality with price. Now, where will you find quality at so low a price as we sell the "BIDDY"? In quality we mean best machine.

Well made, proper theory in its construction, a machine that will do the work it is made for and will last a lifetime with proper care.

The "BIDDY" will hatch more fertile eggs, easiest to operate, requires less oil and lasts longer than any machine we know of. So from every point of view it cannot be excelled.

THE FAITHFUL "BIDDY" DOES GOOD WORK.

Detroit, Mich.

Dear Madam:—I received the pair of Partridge Cochins and am well pleased with them, also "Berry's Biddy," it does good work. Yours truly,
A. H. ALDIS.

SISTER LAWSON HAS RAISED CHICKENS FOR 35 YEARS AND WAS WELL PLEASED WITH THE BIRDS THAT SHE GOT OF ME.

Fairfield, Iowa, 3-14-'12.

Dear Friend:—I received the five Rose Comb Reds yesterday; they are O. K. and I am well satisfied with them. We think they are dandies. We have been raising R. C. Reds for two years and like them better than any chickens I have ever raised except the Rose Comb Black Minorcas. I have been raising chickens for myself for 35 years. I think that I will invest in Runner ducks next year. Will write you later about them. Accept my thanks for fair dealing.
Yours truly,
MRS. W. G. LAWSON.

The "Biddy" does good work in Honolulu, making remarkable hatches for Louis Rosa, a progressive Poultry Fancier there.

How We Make the "Biddy" and Why it is Best

THE CASE.

The improved "BIDDY" is made from the famous air dried California redwood, the best incubator material on earth. It costs more, but it is better. No other lumber for me.

The case is composed of two walls, with a dead air space; each wall is lined with a coat of asbestos, this of course lines the dead air spaces, which insures an even temperature in the machine; no place for admittance of cold air, all joints and bottom corners are mortised and lock nailed, making it impossible for them to come apart or open by usage or rough handling.

The bottom of the case is matched and forced together under great pressure, and held together by mortising and lock nailing with cement coated nails. No crack or holes possible. Everything constructed by the best workmen and highest grade of workmanship.

The top is also put together under pressure and held with a neatly fitted white pine moulding, which adds as a double purpose in holding it together in permanent shape and adding to the appearance of a finished machine. We make our cases better than any other manufacturer we know of.

THE DOORS.

Our incubators are equipped with two doors, an inner glass door, and an outer door made of same material as incubator.

Experience has taught us that eggs hatch much better in the dark than exposed to light, therefore we use an outside door of wood. This extra door also insures a more even temperature along the front of the machine. Our doors are equipped with automatic locks; simply push the door shut and it is locked. No cheap buttons used that are unhandy and liable to come off. Why not get an incubator that is built up-to-date in every respect. The "BIDDY" is one of these.

EGG TRAYS.

Our egg trays are not the flimsy one-piece frame as accompanies a great many machines; but they are constructed of good strong material (white pine) over which is stretched the best grade of wire cloth. As is stated above, these frames are not one piece, but are doubled, to insure against any twisting or sagging of the wire cloth. You know what an inconvenience it is to handle a flimsy twisted tray full of eggs. You will not have this trouble in the "BIDDY."

NURSERY.

The nursery in our incubators is provided with a movable bottom which can be taken out and cleaned. You can also take out all the chicks at once on the tray without disturbing the egg chamber. This bottom has a cloth cover which serves to give the chicks a good footing and prevents spraddle legs. The nursery is four inches in height, which is not too high or too low. Most machines made are to the extreme both ways; either too low so that they can not be properly heated, or not low enough so that the little chicks cannot stand up straight in it.

Every part of the "BIDDY" is so constructed as to add money to the purchaser's income.

THE LAMP.

We have found it necessary to construct our lamps from the very best material. We make these from extra heavy galvanized iron, such as is leak proof and will stand an accidental fall.

Our burners are the best burners made in the United States, a recent patent has made them far superior to anything yet manufactured. You should by all means know what the lamp is like that accompanies a machine you buy, for in the lamp is the power that does the work of the machine.

We use an extra long flue, so that smoking is impossible. Too short a flue is usually the cause of lamps smoking. Our lamps have helped to make the "BIDDY" famous.

HOT WATER HEATER.

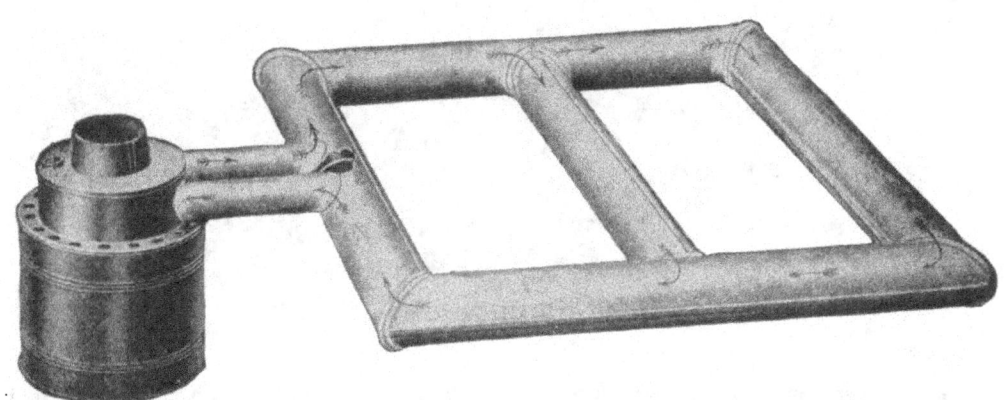

Years of experience have taught us that the heater must be of the best material and made by experienced workmen. A leaky tank causes more grief and dissatisfaction than any one thing about the machine. We have overcome this, by making our tanks from the very best cold rolled copper. We double turn and lock our joints before we put on the solder. The old cheap method of doing this is simply butting the ends together and applying solder. If the solder breaks you have a leak. In our method we make the tank tight before we put on the solder, which doubly insures its strength. These tanks are placed in water and a 25 to 30 pound pressure turned on them. If there is a leak, bubbles will appear on the surface. A tank must withstand this test and prove sound before it is put in a machine.

The pipes make a complete circulation around the inside of the machine, and one pipe comes across the center, connecting with both sides of the circuit. This insures plenty of radiation and even temperature. The pipes enter at two different points in the boiler, one taking hot water from the tank, the other bringing cold water back, thus forming a complete circulation of two independent currents that adds to the heating of the machine. This perfect circulation is found only in the "BIDDY."

PROUD OF BERRY'S BIDDY.

This lady got one of our 120 egg machines and likes it so well that she is thinking of getting a larger one. She favors us with a picture as to the left where she is feeding the chickens with her boys standing beside the "Biddy."

Kahoka, Mo.

Dear Mrs. Berry:—I have been thinking I would write you ever since my incubator hatched the last time, but have just neglected it. I want to tell you that I am proud of my "Biddy." I think that I did well both times that it hatched. I received better results than any neighbors I have heard of. Yours very truly,
MRS. C. E. BROTHERTON.

These cuts are from a photograph of the "BIDDY" we sawed in two from right to left showing the double walls with their inside linings and dead air space. Also shows the position of the heater and the working of the sensitive regulator. The simplest and most scientifically constructed incubator made.

SELF REGULATOR.

The new self regulator attachment on the "BIDDY" is simple, yet durable and effective. I expect all the incubators you have seen the lever works on a little needle fulcrum; but if you should get a "BIDDY" you would see something a little different, an improvement to the machine that can hardly be expressed in words. On our machine the lever works on a knife edge; this insures the little cap on the end of the lever to come squarely down upon the flue, not to one side or the other, as is usually the case in the old way. Our lever is held in position and does not sag to either side.

We have our regulators fixed in such a simple fashion with so small a friction, that with breathing upon the wafer the expansion affects the regulator. No other regulator manufactured moves so easily with the slightest change. Why invest in cheap machines, when you can buy a "BIDDY" as cheap and have a machine you can trust.

VENTILATION.

Building an incubator with the proper ventilation is the most essential part in the building of a machine. A poorly ventilated machine will never hatch the eggs. The ventilation in the "BIDDY" is the very best, and it is such ventilation that will produce moisture enough no matter where the machine is set; so many machines have to be run by the side of a baseburner or a stove that is fired night and day. The "BIDDY" will run any place that you feel safe to set a hen. Our incubators are constructed so that you can keep them by a fire in the day time and without one at night, or you can set it in a place where the temperature is very uncertain. The regulator and system of ventilating will run the machine as it should be run. There is no need of explaining the philosophy of the ventilation in our incubators, space will not permit; but you can rely on it when we say that it is perfect and cannot be improved upon.

You get in the "BIDDY" everything that goes to make one of the most practical machines on the market.

MOISTURE IN THE BIDDY.

The moisture proposition in any machine is one of the most essential. In fact, as much of an essential as the keeping of the proper temperature. The Biddy's ventilation is so arranged that in most latitudes and under most conditions, it is not necessary to apply artificial moisture, but it is hard to build a machine of any kind to overcome all conditions, and for that reason, must make a little explanation herewith which will be of great benefit to all incubator users.

If there is enough humidity in the air, and the outside conditions are favorable, the ventilation will take enough of that air, so that the eggs can extract all the moisture that they want. But these conditions are not the same in different localities. We therefore recommend that a little sprinkling of warm water is given the eggs at each time that you take the eggs out for an airing and turning. Be sure that the sprinkling is done when you are ready to place the eggs back in the machine, so that they will not become exposed to the outside air.

If the weather has been exceptionally warm and dry, it is well in addition to the sprinkling, that you place a little sand tray in the nursery. This sand tray need only be a shallow pie pan, filled with dampened sand. It might also be well under these conditions to dip the eggs at the time they are due to hatch. The moisture keeps the shells soft and pliable, so that there is no danger of the chicks becoming smothered in the shell or unable to break themselves loose at the time of hatching.

There is no incubator on earth that can overcome the weather conditions to such an extent as to give satisfactory hatches at all times without a little moisture application. A hen can hatch the eggs without bothering about sprinkling the eggs simply because she supplies the eggs with moisture from her body. In fact, the heat from her body contains a great deal of moisture, and contains the humidity that we look for in the air, and when it exists, there is no necessity for sprinkling the eggs, because as I have above stated, the incubator will take from the air through its perfect process of ventilation what moisture is necessary.

You will find, by following the above instructions, that more chicks can be hatched, and that they will be larger, healthier, and more vigorous. Always use water about the same temperature as the eggs; do not use too hot or too cold.

OUR EXPERIENCE IS PRACTICAL.

You are all aware that there are many upstarts in the incubator business. At it one season and gone again the next. They are imitators; they imagine that there is lots of money in the business and think that they can make it, too. They know little or nothing about the incubator and poultry business; but make a great bluff by mis-statements, exaggerated representation, and succeed in fooling some into buying. It is all theory and what some one else tells them and not from actual experience. No matter how much a person reads, how much others tell them, how smart or intelligent they may be, they will not know as much or get at the real needs or can make a success of it as one that has had the actual experience.

These sidewalk poultry men and incubator manufacturers cannot sell as good poultry and incubators as one that has practical experience on a farm, raising poultry and using an incubator successfully.

I raised poultry long before there was such a thing as artificial hatching. I grew right up in the business and have the actual experience that has made it a success, so am in a position to offer the best and it will give you the same satisfaction it has me. There is no knowledge that is as good as that gained from actual experience. This should count for much in choosing the "BIDDY" and doing business with me.

BERRY'S "BIDDY" NO. 1—360-EGG CAPACITY.

The above represents the largest machine that we have to offer. It is a machine that will do the work of the smaller machines. It is very often voiced about that the larger machines do not give the proper satisfaction. This might be true of the cheaper machines because they lack radiation, but we spare no expense in making the "BIDDY" just what it should be. We first make a machine that will hatch every fertile egg and then put a reasonable price on it. We don't make our price first, and then have the machine come up to that, and consequently have to cut the quality of the machine down at every turn.

The above capacity is just the right size for any one expecting to make poultry a business or who wishes to make it a good side line. What is the use of being bothered with young chicks all spring and summer? Hatch what you want right early, have them all about the same age; you will find the investment a splendid one.

The price of the above machine naturally would be more than that of any other machine we list, because it is the largest. But stop to figure a little with me and you will find it the cheapest. It is twice as large as the 180 size; what about the price, does it cost twice as much? It is one-third larger than the 240, does it cost one-third more?

Let us settle the incubator and brooder question for you.

```
Complete machine, freight prepaid.....................$24.00
With 200 chick indoor brooder......................... 35.00
With 200 chick outdoor brooder........................ 40.00
```

EVERYTHING SATISFACTORY, AND PROVES A GOOD INVESTMENT.

Dear Mrs. Berry:—   Knapp, Wis., 4-22-'13.

I thought I must answer the letter that I received several days ago. I received the drinking fountains the same day and was very well pleased with them. My ducks have layed 90 eggs all ready. I have sold one setting of eggs and have 5 setting of my own set. The brooder coop arrived about one week ago, but have not put it together yet.

Yours truly,   FLORENCE WATERBURY.

BERRY'S BIDDY NO. 2—240-EGG CAPACITY.

The above is a cut of our new 240 egg incubator. One of our standard and one that we are proud of. This is the most practical size for everybody, and one that will save you money in every respect. We would recommend this machine to all farmers and practical poultry raisers.

Any of our machines in your neighborhood will sell one dozen more. They simply take the lead. The material and workmanship of the above machine is only such as is put in the best machines in the United States, and I am safe in saying that the "BIDDY" leads them all.

If you decide to hatch only a few chicks, you may do so just as easily and successfully in this machine as one of the smaller capacity, by putting in fewer eggs.

We can give you better value for your money in the 240 egg capacity than in any of the smaller machines, for the simple reason that it takes but very little more material or labor to make it than the smaller machines.

This machine is built to hatch the eggs and it has certainly proved its efficiency. You simply cannot lose in getting one of these machines. A trial will convince you.

We pay the freight, something you need not worry over.

PRICE.

Complete machine, freight paid.....................$20.00
With indoor brooder, No. 1........................ 31.50
With outdoor brooder, No. 1....................... 35.50

WOULD NOT SELL FOR TWICE THE COST OF THE BIRD.

Dear Friend:— Postville, Ia., 3-7-'13.
We received our cock alright, and think he is grand. Would not take twice the amount we paid for him. Also the gander that came with him, he was for my husband's brother. He was also well pleased.
I am also going to send you an order for some eggs a little later on. I wish to thank you very much for the book, Advertising Manual, as it was for me.
Yours truly, MRS. MINNIE WILLIS.

BERRY'S "BIDDY" NO. 3—180-EGG CAPACITY.

This is certainly a splendid size and exactly the same model as the 360 and 240 size, made like them in every way, except in the capacity; it has only the one tray while the others have two.

This splendid machine, like all our machines, is backed by hundreds of testimonials, which go to prove that there is none like the "BIDDY."

This is a good size for one that is contemplating going into the chicken business in a small way and does not want a large machine; it uses less oil than any other machine on the market, considering size, and taken all through I would recommend it to most any one.

```
165 EGG MACHINE, freight paid.....................$17.50
With Indoor Brooder, No. 2........................ 26.00
With Outdoor Brooder, No. 2....................... 31.00
```

WE HELP IN A BUSINESS WAY ALL THAT IS POSSIBLE.

Dear Madam:—   New Orleans, La., 1-13-'13.

Received hens this A. M. all O. K. and thanks for the prompt shipment and such fine birds. Anything I can do to improve your business in this locality I will be only too glad to do it.

The first hen you sent me is laying splendidly, and if the four I received today lay as well I will be more than satisfied. With the cock-bird I have them mated to, I expect good results. Wishing you all kinds of success in your undertakings, I am.
                                  Yours truly,   OTTO W. THOMAN.

SO WELL PLEASED. IS ANXIOUS TO RECEIVE MORE BIRDS RIGHT AWAY.

Dear Friend:—   Waterbury, Conn., 3-8-'13.

I received the birds all O. K. and must say that I am very much pleased with them. I thank you for your quick shipment, and will just say that I would like to have two more hens if you can spare them. Please let me know in your next letter how much you want me to send for them. I want to send the money right away.
                                  Yours truly,   F. W. Snyder.

BERRY'S "BIDDY" NO. 4—120-EGG CAPACITY.

The above picture illustrates a very popular size machine and one that I can highly recommend. I believe that I have more people ask about a machine that holds from 100 to 125 eggs than any one other size; of course it depends largely on conditions as to just which size would be the most profitable for you to purchase. All of our machines are made after the same model, each one has the same good qualities of the others. I believe that the above picture shows up the top ventilator a little better than the others, this top ventilator is a new feature and we put it on every machine, it solves the "dead chick in the shell" problem, using this top ventilator according to directions the machine will be supplied with just enough fresh air all the time, so that sufficient moisture can be extracted through natural processes to bring the chicks out strong and healthy.

The matter of choosing the size machine must be left with you, unless you wish to submit to us your circumstances and conditions, and we will be more than pleased to offer suggestions.

We ship all of our machines complete ready for business, except eggs and oil.

PRICE.

Complete Machine, freight paid	$15.00
With Indoor Brooder, No. 3	22.50
With Outdoor Brooder, No. 3	26.50

READ TWO BROTHERS EXPERIENCE.

Dear Madam:—   Smithton, Ill., 5-26-'13.

Received the setting of S. C. Rhode Island Red eggs in good condition, and hatched 10 nice chicks. I am well satisfied with them.

My brother had some shipped from another firm and paid the same price for them, and only hatched 2 poor chicks. I will send you another order in about a month from now.

Yours truly,   ADAM WACHTEL.

SO WELL PLEASED WITH HAMBURGS, WANTS ANOTHER VARIETY.

Dear Madam:—   New Haven, Conn., 4-15-'13.

Your excellent stock of Hamburg chickens arrived in fine condition, and I thank you for such good attention in regard to the quality of the birds. I would like to get a different variety of chickens, the Partridge Plymouth Rocks, one cockerel and a dozen hens. Please quote me your price in your next letter, but I will promise not to be so long in paying for them.

Yours truly,   MRS. H. SUTTON.

"BIDDY" BROODERS WILL RAISE YOUR CHICKS.

Have you ever seen the famous "BIDDY" brooder? Made famous by raising the greatest per cent of chicks put in them. This is what you are after, is it not?

The above is an illustration of three sizes of indoor brooders, 100, 150 and 200 chick size. When you buy our 100 capacity you buy the same size that others offer or claim 150; the same proportion prevailing in other sizes. This you see to start with, gives you a larger machine cheaper.

Many an incubator manufacturer advertises a much cheaper constructed brooder, as either IN-door or OUT-door; but when you are once in possession of your goods you find that you have to set up a stove in some well protected room for them. Not so with the "BIDDY." These brooders are perfect. You can use them outside after March 1st, putting them under some shelter during severe storms. There is surely some kind of a shed on every farm that a brooder can be placed in, and I would not ask for a better thing to raise the chicks than one of these brooders with a little shelter.

We have these famous brooders so well advertised that we thought it unnecessary to give a long, monotonous description of them that would make you tired to wade through it, and you know no more when you are through than when you began. A hint to the wise is sufficient. My guarantee ought to sell every incubator and brooder I can manufacture without a single preliminary sentence.

For the benefit of those who are not acquainted with our machine, allow me to add a few words.

Our brooders are made of white pine and equipped with a hot air system, the only system on earth for a brooder. We have a system of heating installed in our brooders that is not equalled by another brooder manufacturer. I have tried them all. I make these brooders from practical experience, "NOT FROM THEORY." When you buy one of these brooders you buy the brooders that have raised the chicks in our factory and we know what they will do. They are not a cracker box with a cheap tin lamp and a would-be heating system, etc., that you might just as well break up for kindling as to trust your little chicks to. I would rather put my chicks right out in the cold hen house rather than in some of the brooders I have seen.

BERRY'S GOLDEN RULE POULTRY FARM, CLARINDA, IOWA 107

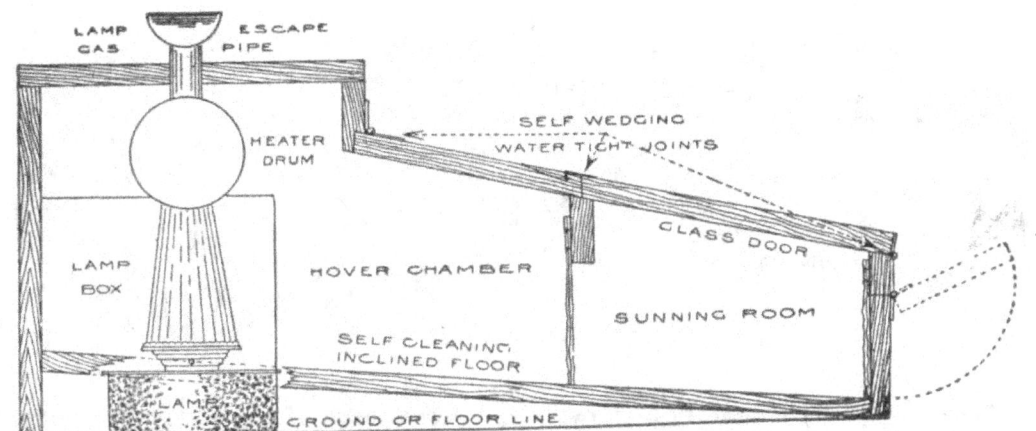

The above shows a cross section of our famous indoor brooder, so that you may see into the working parts and the scientific construction.

Remember that all our brooders are equipped with a wire netting yard, which adds greatly to the usefulness of them.

If you have a brooder house or can use an outbuilding, woodshed, back kitchen, or a smokehouse would do by putting in a south window, I would advise you to get an indoor brooder. My indoor brooders are the best that can be made. Not a bit of cheap material in them. They are made to raise the chickens, and they are brooders in every sense of the word. They are real hen mothers with the capacity of many times that of a hen and they do not fuss around and step on the little fellows, or go trailing out in the wet grass or weeds, or breed mites and lice. So many incubator manufacturers lose sight of the importance of a good brooder; but think of nothing but making a cheap-skate box that is the death of more chicks than they raise and turn people against incubators and brooders.

Not so with me. I know what is the best, and I say it can't be made and sold for as low a price as many of them sell at.

I can make them as cheap as any one can and make them on a closer margin than the people that depend altogether on just the incubator and brooder trade. I sell all the pure blood poultry and eggs that I can raise, which is a good many thousand dollars worth each year.

I tell you that the "BIDDY" brooder is all right and that there is a big value for your money and that it is a machine that will make you money.

If my incubators and brooders were in every home in the United States the chicken industry would be just doubled and the chicken business would be one of the greatest pleasures of the day.

If you wish a brooder that is a BROODER, and one that you can trust your chicks with, the "BIDDY" is the only one that I could recommend as absolutely safe. Get either of the above sizes to match your machine or suit yourself.

We pay the freight. You need not be troubled over excessive charges.

PRICE.

No. 1 Indoor Brooder, complete, freight paid.........$12.00
No. 2 Indoor Brooder, complete, freight paid......... 10.50
No. 3 Indoor Brooder, complete, freight paid......... 9.50

VERY WELL SATISFIED WITH HATCH.

Dear Friend:— Thompson Station, Tenn., 4-15-'13.
As I promised to let you know about my hatch, will say that I have just taken it off. Out of the 175 White Leghorn eggs I got 101, and from the 25 Orpington, 13 chicks. I am well pleased and was not counting on as good a hatch.

Yourstruly, W. T. HALL.

THE "BIDDY" OUTDOOR BROODER.

The above is an illustration showing the three sizes of our outdoor brooders, and when I say OUT-door I mean OUT-door. These brooders are constructed so that they will absolutely keep out the snow and rain. Doors and windows are made to fit so snug that not the least bit of cold can get in them. These brooders are made to be used out in the open air in all kinds of weather.

The corners are iron bound and cannot come apart; the top is tongued and grooved material, insulated or lined with a felting and then covered with a galvanized sheet steel. What do you think of this for a construction? These brooders have two floors, both are tongued and grooved material, and driven up to fit closely.

Our outdoor brooders are equipped with two rooms, a lower and upper floor; the lower room is the hover, or where the little chicks sleep; this is the room where the artificial heat from the lamp is accumulated; the upper room is the sunning room. A felt curtain separates the rooms.

To make a long story short, will say that this is the best brooder manufactured today. Others are realizing it, for we notice imitators on every hand.

All our brooders are so arranged that doors may be removed easily and the rooms cleaned with very little exertion. I had a brooder once that kept me a guessing how in the world to clean it. I hated the thing for this one reason above others.

One other very decisive fault I have found with brooders is the lack of light. Don't you know that a little chick likes plenty of light and they will rather stay outside and freeze to death than go in a dark brooder? I know that you have experienced this very same defect. You possibly did not realize what was the trouble. You simply said, "chicks did not have sense enough to go in out of the cold." The "BIDDY" has plenty of light, the chicks will go in and get warm. They have sense enough to go into a "BIDDY," I assure you.

Remember the construction of our outdoor brooders is such that will insure their safety out in the weather; double floors, triple roof, insulated walls, material and workmanship that is not excelled.

The price of these outdoor brooders is higher than many ask for theirs, but let me tell you, friends, that many sell such brooders as I term indoor brooders, as outdoor brooders. I would not do that; I have manufactured an outdoor brooder that is an OUT-door brooder. The quality is there and absolutely big value for the money and sold at a lower profit than the cheap boxes manufacturers use.

This is a brooder that broods and raises the chicks, and if you do not have a brooder house or suitable building to use the brooder in, our outdoor brooder is absolutely safe and the very best one that is constructed.

Don't let the price worry you. Get one of these brooders and you get something; something that will more than pay for the difference in the first hatch of chickens you take off. You take my word for it. I have experimented with most of the brooders and I say don't buy a cheap brooder. You will be sorry if you do. Mark my word.

Get the best; we offer it in our brooder. Sold under our iron clad guarantee. You take no risk.

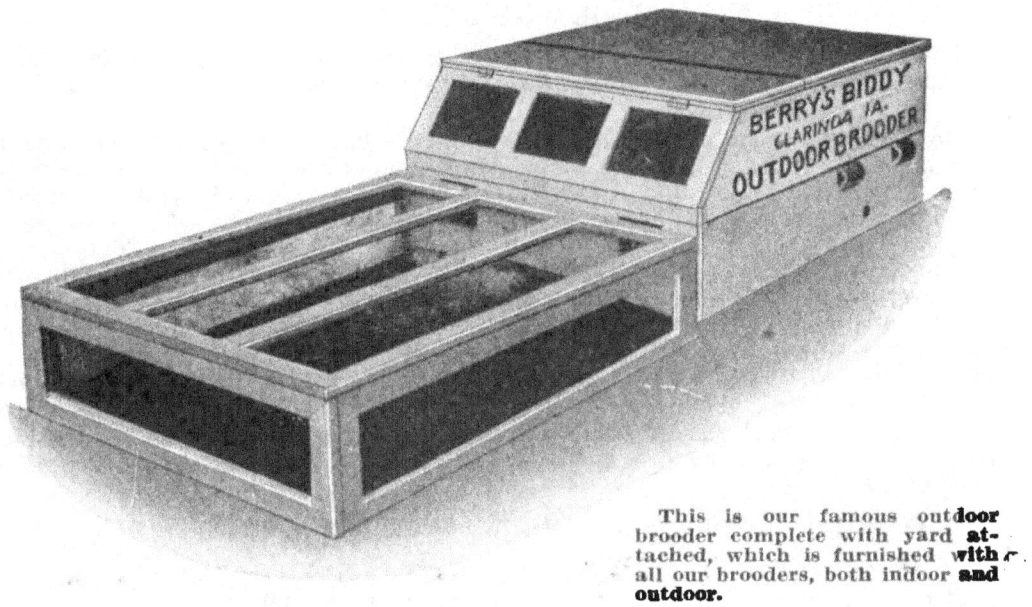

This is our famous outdoor brooder complete with yard attached, which is furnished with all our brooders, both indoor and outdoor.

If you want a brooder better than any other brooder on the market today, you must buy a "BIDDY," the only one of its kind. Buy one and compare it with your neighbors, then you will be convinced. We pay the freight.

PRICE.

No. 1 Outdoor Brooder, complete, freight paid........$17.00
No. 2 Outdoor Brooder, complete, freight paid........ 14.50
No. 3 Outdoor Brooder, complete, freight paid........ 13.00

THE MOST WONDERFUL INCUBATOR MANUFACTURED.
(What a Beginner Can Do.)

Kirksville, Missouri.

Dear Madam:—This is my first season's experience in the poultry business. I did not have the means with which to purchase a lot of fine chickens, so I purchased a "Berry's Biddy" incubator and some eggs. I set the incubator three times and have raised 600 chickens. If I had purchased a brooder instead of trying to make one I could have kept a great many from dying.

Everyone thinks it strange to see so many chickens and not an old hen on the place. I am convinced that there is big money in the chicken business. I am enclosing with this letter two pictures of the most wonderful incubator ever manufactured, "Berry's Biddy."

Yours truly,   W. I. McCLAY.

OUR GUARANTEE.

There are many who write flowery catalogs and make strong assertions in their advertisements as to their guarantee; but you cannot find a positive guarantee anywhere in their catalog.

Now, I make a strong Iron Clad Guarantee that means something—and it is in plain sight in a conspicuous place. It is a guarantee that means just what it says. It means a guarantee of a square deal. I would not need to make it to the many of my old customers who have bought stock and eggs from Berry's Poultry Farm as they know by actual experience that I give a square deal—big value for the money; but this book goes into the hands of many new friends for the first time, and it is human nature to be conservative, a little slow of getting acquainted through correspondence and printed book; so I make such a strong guarantee and give such well known and reliable references that you can not help but know I will give you a square deal, as you take no risk in getting acquainted and doing business with me.

So let us do business together and get started in this great chicken game. I have the chickens, incubators and eggs; you want to get started, so come on with your order.

WHY YOU SHOULD BUY BERRY'S "BIDDY" INCUBATORS AND BROODERS.

Because they are built by people who have had actual experience in artificial hatching and rearing of chicks. Therefore, they are the result of experience rather than theory.

Because they are made of the very best material throughout.

Because they are sold at the lowest possible price consistent with good business principles.

Because the "BIDDY" incubator will hatch more and stronger chicks than any other machine on the market.

Because the regulating and heating device has many improvements that save oil, time and work.

Because they are so constructed the attention they do require is only a matter of a few minutes each day, and the instructions are so simple and the work so easy that a child will have no trouble operating one of them.

Because the "BIDDY" brooders are brooders in the broadest sense of the term, and not death traps in which the lives of your young chicks are sacrificed.

Because the "BIDDY" brooders are made with the same care devoted to the "BIDDY" incubator and has a regulator that controls the heat of the nursery where the chicks are brooded, and is also provided with a good thermometer and you don't have to guess at the temperature to which the chicks are subjected.

Because the ventilating system on the "BIDDY" brooders is constructed on scientific principles, and the chicks are provided with warm, fresh air at all times. In fact there is no chance for the air to become foul.

Because these machines are sold under our IRON CLAD guarantee, and you risk nothing. They must meet the claims made for them or we get the machine back and you get your money.

IN CONCLUSION.

We could give fifty more reasons backed by strong argument why you should buy "BIDDY" incubators and brooders, but will sum them all up in one which is strong enough to make you decide that you should buy a "BIDDY":

The "BIDDY" are the Cheapest and Best Machines made, taking everything into consideration. We have proven this statement ourselves and know it is so. Buy one and let us prove it to you.

If we can't it will not cost you a cent.

THE INSURANCE FIRELESS BROODER.

Best thing yet. The right way to raise chicks—made of heavy Galvanized Iron throughout.

This is little Thelma, 7 years old, daughter of Mr. and Mrs. R. L. Kirby, of Beatrice, Neb., Berry's "Biddy" Incubator and Fireless Brooder. Now, is not that a very pretty sight and sure prosperous? I think Thelma had a dress made for 4th of July celebration.

Beatrice, Neb., June 11, 1912.

Dear Madam:—I bought one of your Insurance Fireless Brooders this spring, will say it sure is a treasure—paid for itself the first hatch, because I never lost a single chick. I like it better than any brooder I ever used and I have tried a number. I felt so proud of it I just had some photos taken with my little girl, Thelma, 7 years old; so I will send you one. I have 220 S. C. Reds and raised them all in the Fireless Brooder.

Yours very truly,

R. L. KIRBY.

See the above picture of one of these fireless brooders in actual use. A photograph taken by one of my customers. This speaks volumes.

The Fireless Brooder is so constructed you can raise your chickens without a lamp or artificial heat.

No cost to run it.

The chicks will be stronger and healthier.

No danger of having a fire from lamp or heaters.

It will last a life time.

In short, it will raise more and healthier chickens at a much less cost than any other way of raising chickens.

I use them on Berry's Golden Rule Poultry Farm and know they are good.

They beat anything yet offered to raise chickens and the best thing is their durability and low cost.

I guarantee them to do just what I claim for them.

It is made of the best galvanized steel. It protects your chicks from rats, and lice will not breed in them. It is made with a hover and does not require a lamp—it is so constructed the chicks will furnish their own heat. They can be used in and out of doors to accommodate the weather. They are easy to clean, the top slides off. The bottom of the feed room and hover are made seperate and can be taken out one at a time to clean. The partition that separates the hover from the feed room can be taken out to accommodate the growing chicks. It is no small cheese box brooder.

No. 1 for 50 chicks: The hover measures 13 by 25 inches or 325 square inches; feed room, 16 by 25 inches or 400 square inches. The run or yard attached 25 by 32 inches or 800 square inches. The Brooder when set up measures 2 ft. 1 inch wide and 5 feet long. **The price of this brooder complete with wire run, $5.25.**

No. 2 for 100 chicks. Hover 20 by 25 inches—500 square inches; feed room 22 by 25 inches—550 square inches; run or yard, 25 by 42 inches—1050 square inches; set up complete, 2 ft. 1 inch wide and 7 feet long. **Price, complete with wire run, $7.25.**

This shows a splendid picture of our latest Fireless Brooder. The top is removed so that you can see the hover in position. The roof on this brooder slopes only one way, while the roof of brooders No. 1 and No. 2 is a hip roof.

This is a new Fireless Brooder that we are making this year; it is practically the same as our No. 1 and No. 2 with the exception that it is made without a feed room, but so many of my customers have such facilities that it is not necessary to have a feed room in the brooder, they can feed their little chicks on the outside of the brooder proper, yet have them in shelter. For these customers we offer this brooder, which will hold the capacity that we claim for them. This brooder is considerable cheaper and is certainly going to be a very popular brooder for your little chicks, once you try them you will never try to raise a brood without one or more.

No. 3, 50 chick size, coop proper is 16 by 25 inches, yard or run way is 24 by 28 inches. Brooder fully equipped and ready for business.

Price, complete with wire run or yard, $3.75; without wire run or yard, $2.75.

No. 4, 100 chick size, brooder proper, 24 by 30 inches; yard or run way is 24 by 36 inches. Brooder fully equipped and ready for business.

Price, complete with wire run or yard, $5.00; without wire run or yard, $3.75.

THE INSURANCE NEST BOX.

Why not furnish your chicken house with Galvanized Nests?

Will not breed lice. Your hens can lay and set in them and not be tormented with lice They will last as long as you want them. Made in sections of three and easy to handle. You can put them on a shelf, hang them up or set them on the ground. They are furnished with sliding doors so that the hen may be shut on or off the nest. The Insurance Nest Box is a good investment.

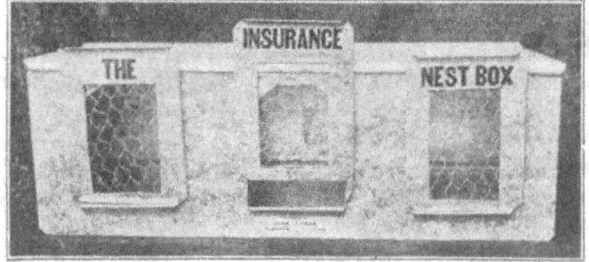

Price, $2.50.

In Use with Yard.

Insurance Brood Coops

Insures your little chicks against rats, minks, weasles and vermin of all kinds as well as lice and mites.

Say, have you not had some very disappointing experiences with little chickens, after they are hatched? A good many claim that it is harder to raise chickens than to hatch them. It certainly is unless you are fixed.

Now, a friend of mine has invented and patented a coop constructed of heavy galvanized iron that will last a lifetime and it is the best thing that has ever been gotten out for raising little chicks. It is successful, it is sanitary, it is durable, it is rain proof, it is light and easy to move about, easy to clean and a whole lot the best thing that has ever been invented or put on the market for a hen and little chicks.

You can also use it as a fireless brooder by putting in an adjustable hover, and a gallon or gallon and a half jug of hot water will keep it warm for twelve hours at a time and during the month of June and later it will require no heat only on damp, chilly days.

It is so constructed that it can be taken apart very quickly. We ship it in a knock-down and can be put together in five minutes. When you get through with it at the end of the season it can be taken apart and put away, occupying very little space until the next season, or you can use it as a nest, placing it in your hen house; it is fine for that purpose. The mites and lice will not bother it and it is easy to clean and makes the best nest there is.

Construction. It is made of heavy galvanized iron with no seams exposed or no bolts or rivets. It is put together with a peculiar lock joint that is strong and efficient. It has the best ventilation arrangement of any metal coop on the market. The bottom is so arranged that it can be taken out in a jiffy and cleaned so you do not have to take it apart to get the bottom out, as some other metal coops. In fact it is differently constructed from any other metal coop and is fully covered by a patent which in applying for, the patent attorney advises us that this is the latest patent device in this line.

There is a covered brooder or run-way that can be attached to this coop by four small bolts. This is fine for the little chicks until they get to be a week or so old as you can move it about every day and have a nice fresh place for them and let them exercise themselves in the sunlight. A little later you can keep the hen shut up in the run-way and open the little chick door and the little chicks can run at will.

This covered run-way is a fine thing. It is proof against all kinds of vermin unless they dig under it. In warm summer weather the coop door can be left entirely open, as the run-way will keep out the vermin and thus giving them a lot of ventilation and keeping them cool during hot nights. However, the coop is successful without the yard, but is better with them and you should have at least a few of the complete coops and yards.

This is a splendid thing to put brooder chicks in after they are three to four weeks old in early spring, and you can raise them right from the start after the weather gets reasonably warm, very successfully.

Now I tell you, if you want the best thing that has ever been invented for raising little chicks, get one or more of these. The price is moderate, but its value cannot be estimated and there is not a chicken raiser but what will save many times over in the course of a year as the loss of one night from a weasel,

Closed and without **Yard**.

rat or mink would pay for a good many coops and leave a large amount of pin money besides.

The cost is nothing compared with the value. I am safe in saying that one of them will last for forty years. Galvanized iron, as you know, is very substantial; it does not rust or rot and five cents per year will pay for it; so what is five cents when you compare it with what it will cost you to make a wooden coop that will rot out in a few years and will not be so sanitary nor so good in any way. These are such a good thing that I am expecting to sell a good many thousand of them this year, as everyone that hatches chickens stands in their own light if they do not secure some of these Insurance Brood Coops.

PRICES.

No. 1—For chickens and ducks; 16 by 24 inches; 16 inches high in front, 12 inches in back.

1 Coop, $1.75; 2 for $3.35; 4 for $6.25; 6 or more, $9.50.

With run-way or yard:

1 Coop, $3.50; 2 for $6.75; 4 for $13.25; 6 or more, $19.50.

No. 2—For turkeys and geese; 25 by 30 inches; 22 inches high in front and 19 inches in back.

1 Coop, $3.00; 2 for $5.75; 4 for $11.50; 6 for $17.25.

With run-way or yard:

1 Coop, $5.00; 2 for $9.75; 4 for $19.50; 6 for $29.00.

Either size can be used for brooder chickens. No. 2 is the largest size and will hold the most, and is almost a small colony house. They can be used for coops after the plan of the Philo system.

HOG AND POULTRY TROUGH

This is a splendid trough; I wish that you could see it. This illustration will give you a pretty good idea of what it is like. It is constructed so that it will last a life time. The hog trough is of course the most substantial; it is in fact built so strong that a loaded wagon could not make an impression if allowed to run over it. I do want you to have one of these troughs, and I am making the price exceptionally close.

The poultry trough makes a splendid good trough for ducks to feed their mash in, also makes a very good watering trough for grown ducks and chickens. Buy one and be convinced that it is just what you want.

Price

Hog trough, 8 ft. long, $5.50.

Poultry trough, 5 ft. long, $2.25; 2½ ft. long, $1.50.

Berry's Fireless Cookers

No. 3 Compartment Fireless Cooker, $9.75.   No. 2 Compartment Fireless Cooker, $8.50.

I suppose you will wonder what Fireless Cookers have to do with the poultry business. It hasn't any direct connection, but they are such a splendid good thing that I just had to offer my customers the best one that is made at a reasonable price. It is a cooker that is scientifically and better made, will last longer and do better work than any one that is offered.

Now, I know, as I have tried a number of them. I sent for one very extensively advertised and did not have it a week until it came apart in the joints, so of course, not being tight, it lost much of its usefulness. The next one warped all out of shape and made openings in it that gave ventilation, which rendered it useless. The utensils were made of tin or cheap granite ware and could not last over six months at very best; simply racket store goods and offered to sell only. So it went; they either sprung open, warped or the vessels gave out in a short time.

There is nothing mysterious about Fireless Cookers. The entire plan is to retain the heat. Water turns to steam at a certain temperature, and if you boil it a month it would get no hotter; the Fireless Cooker simply retains the the heat and finishes the cooking.

It saves fuel, it saves worry and working over a hot stove, and the natural flavors are retained. Every one that has used them will testify that food cooked in a Fireless Cooker is much superior over the old way.

There is no machinery, no regulating, no learning "How." It simply keeps the heat in—absolutely a non-conductor of heat.

Would you not buy a stove that cooks without fuel? That is what the Berry will do. Don't you see how quickly such a stove will pay for itself?

Prepare any food to be cooked in the usual way. Bring it to the boiling or roasting point over your cook stove, place it in one of the air tight receptacles of the Fireless and it will continue to cook as long as it remains there.

Then as it requires no watching, your time is your own. Cooking is off your mind until you are ready to serve.

The Berry's Cooker will boil, roast, bake or fry. It is exceptionally good for roasting meat and fowls. With this method of roasting, there is no attention given to the food, and no basting, no watching, no trouble of any kind, as no evaporation takes place. You can make the finest soup by taking and starting it, and then putting it in the Cooker and it will be piping hot, ready to serve the next meal.

And such delicious pies it will bake. There is no other process or any other way that you can produce such a fine pie or biscuits as in the Berry's Fireless Cookers. You can make better coffee and Postum in it than any other way.

It will brown the meat of fowls and the gravy will be there all ready to serve and without further fire or attention. You may start the meal in the Berry's Fireless Cooker, leave the house and spend the day calling, shopping, or in any other way you wish to, and know that when you return in the evening, the meal is ready to put on the table.

A book full of recipes and complete instructions goes with every Cooker. Don't think you will not have success with it because you have never seen one.

Description.

Wood frame, top and bottom; sides covered with Sheet Steel highly finished, trimmed at top with wood moulding, grained in quarter cut oak finish.

The top is covered with Aluminoide, a non-rusting metal. Each compartment is lined with galvanized Steel and furnished with galvanized Steel hood, which fits down into the compartment completely surrounding the Vessel. 1 set (2 only) Metal Radiators for large Roasting and Baking compartment and one only Small Metal Radiator for each of the small Boiling compartments. 1 key for handling radiators.

No. 2 is furnished with one four- and one eight-quart solid aluminum utensils with a patent locking device.

No. 3 is furnished with two four-quart and one eight-quart solid aluminum utensils, with the patent locking device to hold the cover down.

The cookers are packed with mineral wool with the exception of the corners, which are padded with excelsior.

The "Berry" is made of the very best material throughout and in a thoroughly workman-like manner and is fully guaranteed. Compare this with some of the thin wooden boxes with shavings or excelsior packing and racket store tin or cheap granite ware. You can buy such utensils for ten cents; but if you know anything about the price of aluminum you will know that the latter costs a great deal more, in fact there is no comparison; but aluminum will last a lifetime, while the other will last at best six months. It is covered with iron and can't work or pull apart at the joints and seams.

The mineral wool packing that I use is something very expensive, but it adds to the superiority of the cooker and I use it as a packer, even if it does reduce my profit. I have got this cooker on a rather close margin now, as it costs a great deal to construct one like the "Berry," but I want everyone that buys one to be able to say in 20, 30 or 40 years hence, that Berry's Cookers are the best and most substantial thing ever put in the house, and that you can say that it does just as good work after so long a use as it did the day you received it.

It will be a sweeping advertisement for all time to come and then I want to keep up the standard I have set: "A SQUARE DEAL FOR EVERYONE AND BIG VALUE FOR THE MONEY."

I will make the statement and I will back it up with SHOW ME proof.

The Berry Fireless Cooker will:

Cook 80% of the family food.
Save 80% of the fuel bills.
Save 80% time and trouble.
Food 100% better cooked.
Tastes 100% better.
Is 100% healthier.

Then buy a BERRY FIRELESS COOKER and be happy.

PRICE.

No. 2 Berry Fireless Cooker, two compartments, made as before described and fully guaranteed to be the best Fireless Cooker made. Complete with recipe book, $8.50 F. O. B. factory.

No. 3 Berry Fireless Cooker, three compartments, made as before described and fully guaranteed to be the best Fireless Cooker made. Complete with recipe book, $9.75 F. O. B. factory.

The cheapest and best Cooker on the Market.

THE WAY WE SELL BERRY'S FIRELESS COOKERS.

We are just introducing the Fireless Cooker. We want to help you out all that we possibly can and want to place one of our Cookers into every neighborhood, that it is possible for us to introduce it. And we have therefore arranged our clerical work in such a way so that we can sell this article on the installment plan. It is the only thing in our book that we list that it is possible for us to do this. If you are not able to pay the price down that we ask, and if you want one of these Cookers very badly, we will only be too glad to let you have it with a remittance of one-half down, the balance to be paid in two installments of one month apart. If you can pay cash for the Cooker we will allow you a 10% discount, which means that we will let you have the No. 2 compartment 85 cents cheaper than it is listed, and the No. 3 compartment $1.00 cheaper. I believe that this is making you a good inducement, and I hope that it will help you out.

Strawberry Plants
The King of Small Fruits

I suppose some will think it rather peculiar to offer strawberry plants in a Poultry Book by a Poultry Firm, but I don't think there is anything strange about it. We work them in connection with our poultry farm and are found to be so well adapted and such a very profitable crop that I just give my friends a chance to get into the very best combination there is for profit and pleasure.

I do not believe there is anything that goes quite so well with the chicken business as the raising of strawberries. It is admirably adapted in every way to work well with the chicken business to the best advantage. The demand for delicious fruit is universal, the market is scarcely ever over done and the demand so great that good prices prevail, which nets splendid profits. They are a crop that is easy to raise, require only a small plot of land and are admirably adapted to a great many people. They come at a time when business is a little slack in the poultry line and can be worked in as a side line to a good advantage.

Everyone that has as much ground as a blanket will cover should grow strawberries for family use at least; they are a delicious fruit and universally liked, so easily and cheaply raised; it is simply negligence not to embrace your opportunity to grow them. The profits from growing them commercially are very great and you can grow a fourth, a half or an acre or two and the money returns will certainly surprise you. The prices usually paid for strawberries are very good compared to the amount you can raise to the acre. Strawberries can be grown with profit in every state in the Union.

I just recall an instance of a family living in an adjoining state, the husband died a few years ago, leaving the wife and three small boys; they were in the chicken business in a small way and were just getting a good start with things coming along pretty well in that line. The man had also set out one-half acre of strawberries before his death; it was in the spring time and the wife began the task of supporting the family. The oldest boy was fourteen and the youngest 7 years. With the aid of these little fellows the widow put out another half acre and gathered and marketed the berries. It proved very profitable and she was able to start the boys to school that fall. The following season the berries netted her $488.00, which helped out the chicken profits and placed her in a position to live well, enjoy many of the luxuries of life, educate and do well for her family of boys.

CULTURE.

I recommend early planting, the earlier the better in this latitude, which comes in March to May.

The ground should be in good condition, fall plowing is best. Mark the rows three to four feet apart and set the plants 12 to 18 inches apart in the row, this for our climate and field culture; for garden and special culture you can plant them twice as thick, rows 18 to 24 inches apart and plants 6 to 12 inches apart in the row. In the South, especially in Florida, the fields are planted as thick as above stated in garden culture, as they use considerable fertilizer and the plants bear from three to six months after they are set out and do not have a chance to thicken up as ours do in the North. In planting use a dibble, a tool made especially for plant setting, or a spade; a man with a spade and a box with the plants with a little practice can soon set a great many plants in a day. Care should be taken that the roots point downwards and not allowed to bend back or curl up when dropping them in the hole made by the spade, as the plants should be dropped just as the spade is being removed from the earth. Cultivate often; we have a five-tooth cultivator we use a great deal, also a thirteen-tooth adjustable harrow. If the ground is reasonably clean weeds will not bother to any great extent, but the secret is cultivate often and thoroughly. After the ground freezes, when it is thought it is permanent, mulch with some litter, with anything that is convenient —an old straw stack, manure from horse stables, wild hay, or tame hay without any seed in it, as great care must be taken that no seed is allowed to get into the strawberry beds; don't use fresh straw stacks. I saw a neighbor spend $25.00 in getting his acre and one-half of strawberries cleaned from a crop of wheat that sprung up early in the spring from seed that was in the straw he mulched the plants with the fall before.

BEST VARIETIES TO USE.

This question is a good deal like answering the question that is asked me a hundred times in the season as to what varieties of chickens are best. There are some varieties that are adapted to only the far South, while some varieties are adapted only to the North, but we have chosen such varieties as are best adapted for all localities. The Senator Dunlap is the most universally popular for an all-purpose one-plant variety, but for big profits, quality of berries, hardiness to resist frost, drouth and wet weather, our strawberry expert after 26 years of experience finds that the Warfield and Bederwood stand at the head of the list. However, I would not say we have all the best varieties.

HOW MANY TO SET.

A great many people will be undecided as to the number of plants they should set out. Of course, this depends upon your circumstances and what you aim to do along this line. For an ordinary family one to three hundred plants are sufficient to make a good supply, but remember, a few surplus berries is a good thing to have and they are easy to sell. For a small family, 50 to 150 will give sufficient, and for commercial use from one-quarter of an acre up, increasing your plantation as your experience develops, until you have a good business of your own and you can retire from the profits.

VARIETIES.

The sorts marked "P" given herewith are imperfectly flowered and should have a perfectly flowered sort planted in adjoining rows. Varieties marked "S" are perfectly flowered and do not require any other varieties to fertilize or pollenize them, and can be used in connection with imperfect sorts.

In planting for commercial fruit we recommend two rows of Warfield and one of Bederwood.

Senator Dunlap (S)—Medium to late. A great many regard this strawberry as the best one of all varieties and has won great popularity, has long fruiting season, growing immense quantity of large, handsome berries of most excellent quality. They are a general favorite in the market everywhere. The color of the berry is a rich dark red with a glossy finish, shading to dark scarlet on the under side. It has bright yellow colored seed which give it a very rich appearance. It is noted for uniformity in size and shape, and its very delicate flavor and the sureness of its crop make it a general favorite and in every respect ideal. You cannot go wrong to order part of your crop from Senator Dunlap.

Crescent (P)—Early variety. This is one of the old standbys and is certainly a good one and very dear to the hearts of strawberry growers. Very productive and of splendid quality, and is a sure bearer.

Klondike (S)—Medium early. This variety is noted for its very large yield of beautiful berries, uniform in shape and rich blood red color, this color extending through the entire flesh. It has a mild, delicious flavor unlike any other variety and is full of juice. It has tall foliage, with a medium size, light green leaf, this foliage acts as a shade to keep the hot sun from drying out and injuring the berries during a dry season. It is a splendid variety for the South, but is very valuable for all latitudes.

Baldwin's Pride (S)—A wonderful new variety of genuine merit, originated in Michigan, and it has proven its great worth everywhere. It is adapted to all soils, climates and conditions where any strawberry will grow. It is medium early, and bears well into the summer, and holds out with the latest of them. It comes in just with the Senator Dunlap. It is very productive, dark glossy red throughout, of good size, and is a splendid handler and good shipper. It is a perfect blossom and a strong fertilizer with very vigorous plant and vining system, and lots of strong fruit stems. Order some of them by all means.

Baldwin's Pride.

Aroma (S)—Late variety. This is one of the newer sorts and has become very popular as a commercial berry and is a variety which is most exclusively grown in the South and Southwest. Its appearance in a box is most attractive. The prominent yellow seeds on a background of a deep dark red heightens the effect and makes it very attractive. The flavor of the fruit is deliciously aromatic and very rich, the flesh being smooth and solid; one of the firmest and best shipping berries yet ever introduced.

Lovett (S)—Early variety. This berry is rather large, of a deep crimson color. The seeds are a bright yellow, extending well out from the service, and the color effect is rich and pleasing. The flesh is dark red, flavor rich and juicy, possessing just sufficient tartness to make it a splendid canner. As a shipper and market berry Lovett stands very high, as it holds its brightness and gloss as long as any other variety.

Warfield (P)—Early variety. This is an old sort as it stood the time and kept right near to the centre of the hearts of all berry growers. It has always ranked high among the varieties particularly adapted to canning and shipping, the fruit retains after canning its rich fine red, and its delicious flavor is perhaps superior to any other variety. With all these good qualities it is an extraordinarily prolific bearer. Its fruiting season exceedingly long and there is a remarkable uniformity of quality both in early and late yields. Don't fail to try some of them this year.

Perfection (S)—The best strawberry on earth. This is a new variety, and it has been fully tested and proves that it is the peer of all strawberries grown, and is simply perfection in strawberry culture. Perfection is a variety we are pushing and introducing after fully demonstrating its great worth and quality. It has made a wonderful record on our farm. I believe if you plant a field of these it will make you rich; and a small patch, a lot of money, or just a family bed of 10 or 200 plants, will make all you and your neighbors can eat.

Description.—Perfection is very productive, extra large in size, of a beautiful dark lustrous red, with flesh of the same color, all through. It is perfection in quality, fine flavor, and is simply delicious. It has dark green foliage, with strong plant and root system, and with abundant fruit stems. It is a splendid handler and fine shipper.

Price, prepaid: 25 plants for 50c, 50 for 85c, 100 for $1.50.

Your expense: 100 plants for $1.25, 200 for $2.00, 500 for $3.50, 1,000 for $6.00.

Glen Mary (S)—Medium early. This shipping berry is very popular on account of its ability to handle well, of its fine appearance and unusually high flavor; it is a very popular variety. Its yield is very prolific, large, dark red color, with prominent seeds of a bright yellow, making it a very attractive and very popular berry, which seems to increase every year.

Perfection.

Bederwood (S)—Medium early. This variety produces a delicate crimson berry with glossy surface and deep yellow seeds with fruit medium size. The flesh is red, shading down to a rich cream near the heart. Its richness of color, delicate flavor and fine quality makes it one of the most popular varieties with the high-class trade. It brings the big money in the best market. It is exceedingly productive, an excellent shipper, and is popular with commercial shippers everywhere. It has an extremely long blooming season and like the Lovett is an excellent pollenizer for pistillate varieties. This has been a popular berry for the past 25 years, and each year it grows stronger in the esteem of strawberry growers.

Wilfley's Wonderful (P)—Medium early. This is a new sort just introduced; the plants are large, with a good root system; the fruit is very large, of uniform size and very fine quality. Color is a dark rich red which extends through the fruit; it is very firm and solid and an excellent shipper. The blossom is very strong and is a first-class frost resister. This is a sort that I believe is going to become one of the most popular of all varieties and some day will head the list. It is a berry well worth trying.

Price of Strawberry Plants
All Varieties Except Perfection.

Prepaid, by mail or express: 25 plants, 35 cents; 50 plants, 60 cents; 100 plants, $1.00; 200 plants, $1.50.

By express or freight, your expense: 100 plants, 75 cents; 200 plants, $1.35; 300 plants, $1.60; 500 plants, $2.25; 1,000 plants, $3.50.

Ask for special prices in larger quantities.

ASSORTED COLLECTIONS FOR BUSY PEOPLE.

Collections that are best adapted for garden and family use.

Family Collection No. 1.—100 plants, made up of Perfection, Senator Dunlap, Lovett, Bederwood. Charges prepaid, $1.00.

Family Collection No. 2.—200 plants made up of Perfection, Senator Dunlap, Warfield, Bederwood, Lovett and Wilfley's Wonderful. Charges prepaid, $1.75.

Family Collection No. 3.—300 plants, made up of Glen Mary, Senator Dunlap, Warfield, Bederwood, Lovett, Perfection, Baldwin and Pride. Charges prepaid, $2.25; your expense, $2.00.

No changes from above varieties.

Don't fail to ask for special price in large amounts. I will meet any prices, quality of plants considered; but no one has better or surer plants, as all are fully guaranteed.

Good Seeds

Mr. Berry has been in the seed business longer than I have been in the poultry business, and has established a very large and successful business. This has been done on the square deal methods of putting quality before profit and guaranteeing all the seeds that they handle. I know there are a lot of my friends that are interested in seeds and it will be to your interest to know where you can get good, reliable, guaranteed seeds at such prices that will save you money. Good crops come from good seeds well sown at the proper time, combined with good weather and intelligent working of the soil.

In ordering, do not get the two firms confused, as the poultry farm and seed firm are two distinct businesses, and get their mail at two different offices. You can save time by addressing each firm separately, or at least keep the business of each concern on different sheets.

This is our boy Ernest, and one of Mr. Berry's Seed Pumpkins.

GARDEN, FIELD AND FLOWER SEEDS grown in Iowa, the garden state of the Union—whose soil, climate and sunshine all conspire to produce Garden, Field and Flower Seeds of the finest quality and of the highest germinating power in the world.

For the Farmer we have seed corn that will increase the yield and quality to an astonishing degree. We sell it either in the ear or shelled. **Rust-Proof Oats, Early Fife Wheat** that is a sure grower, Dwarf Essex Rape, Spelts, beardless and bearded **Barley**, Cane and **Kaffir Corn** which produces abundant feed; **Seed Potatoes** that can't be beat; **Artichokes**, the great hog food; fine, clean **Grass Seeds** for field or lawn. Our Aristocratic **Lawn Grass** is brand new and will produce a velvet sward fit for the front yard of a king.

For the Gardener we have Garden Seeds that are true to name, of excellent varieties and perfectly clean of weed or other foreign seeds.

For the **Ladies** we have the finest selection of varieties and the freshest **Flower Seeds**. Also Flower Bulbs, among them our celebrated **Paradise Dahlia**, one of the most beautiful of flowers; also the new **Jerusalem Rose**, a wonder among flowers—the most prolific and beautiful of the rose family.

We either grow or directly superintend the growing of all our garden, field and flower seeds and are in position to **absolutely guarantee** the quality, freshness, name and growing power of the seeds we sell.

Our handsome new catalog contains full descriptive information on a special proposition to seed buyers this year. We will send it free on request.

We guarantee all our Seeds. Your order will be appreciated.

A. A. BERRY SEED COMPANY, Box 202, Clarinda, Iowa.

Berry's Poultry Supplies

Who would know more about the needs of poultry than one who was raised on a farm where poultry raising was well looked after and one who had always taken an interest and experimented with the feed and needs of poultry?

That's us.

If there is one thing that we know about more than another, it is poultry supplies, and especially our Berry's Chick and Egg Foods, Lice Killer, Insect Powder and Cholera and Roup Cure, which are undoubtedly the very best there is. They are highly recommended by thousands of pleased customers. The goods are all right and we guarantee them. The prices speak for themselves and we guarantee quality.

BERRY'S EGG FOOD.

Makes lots of eggs. Sold under a positive guarantee. Oftimes you wonder why it is that you are getting no eggs and your neighbors are continually gathering a good supply. It is simply because your hens are out of laying condition; although apparently in as fine a condition as your neighbors. Your hens need something to strengthen and invigorate those organs that have to do with the making up of the eggs; your hens also need the substances that go to make up the eggs; you will find all these elements in our egg food; a few feeds will convince you that your neighbor's hens won't be in it for egg production.

Remember that this egg food is sold under a positive guarantee.

Price, 50 lb. sack, $1.75; 100 lb. sack, $2.50; bags free; F. O. B. Clarinda.

BIRDS SO GOOD THAT MORE ORDERS FOLLOW FROM PLEASED CUSTOMERS

Harlowton, Mont.

Dear Friend:—I write you asking if I can get eggs of the three breeds that we got of you last fall that are not related to them. The chickens you sent are certainly fine and I thank you for sending such nice ones. You shall have more orders from us later on.

Your friend,
MRS. GEO. SELLECK.

Save the chicks—no more corn meal and water; no more bowel trouble; no more dead chicks, but fine, healthy, hearty chicks, is the result of feeding Berry's Chick Food and following the directions as to caring for and raising them. Why go to the trouble of hatching out a lot of chicks and have them all die of bowel trouble? What is the use of killing your chicks by the old fashioned corn meal and water diet when you can procure Berry's Chick Food for but little more than the corn meal? Corn meal, sloppy diet, has killed more chicks than it has kept alive. Such feed and methods are back numbers. Berry's Chick Food has revolutionized the chicken raising business; is absolutely the best combination of grain, seed and animal matter that has ever been discovered or mixed together.

We use nothing but the sweetest of grains, seeds, meat quality, etc., the combination, as every one knows, is the best thing for stomach troubles. Eminent physicians have long prescribed it for man and beast. Bowel trouble, which kills millions of chickens every year, is caused by a diet of raw corn meal and water, or any kind of improper feed. Many so-called chick food mixtures are frauds and are made only to sell. Our sweet grains, seeds, meat scraps, cracked bone, grit and herbs sweeten the stomach to perform the operation. It is highly nutritive and will cure all bowel ailments, cholera and other troubles that affect young chicks. Berry's Chick Food is fed dry, in the natural state, which is a sensible and natural way and which gives such excellent results. Sold under positive guarantee as to results, and no one can put up a better argument than this, or anything that would be stronger.

Price, 50 lb. sack, $1.50; 100 lb. sack, $2.25; F. O. B. Clarinda, Iowa.

BERRY'S ADJUSTABLE LEG BAND.

They are made of aluminum and are the best band made. They are neat, strong and durable, easily and quickly put on, and this band has a double lock, making it much more secure than the single clinch. In fact,

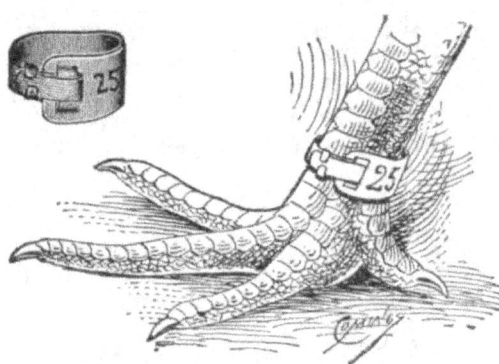

BERRY'S ADJUSTABLE ROCK LEG BANDS.

Its coming off is impossible. This is one of the best features, since a band with any possible chance of coming off is worthless.

Price, by mail, postpaid, numbered with running numbers from 1 up: 12 for 15c; 25 for 30c; 50 for 50c; 100 for 75c; 200 for $1.50; 500 for $3.50; 1,000 for $6.50.

BEST BIRDS IN THE CITY.

New Orleans, La.
Berry's Poultry Farm, Clarinda, Iowa.
Gentlemen:—The Rhode Island Reds have arrived and I desire to compliment you on having sent me the finest lot in the city.
Yours truly, R. G. DUGUE.

BERRY'S AUTOMATIC ALUMINUM DRINKING FOUNTAINS.

The newest and best thing yet discovered.

No chicken yard is complete without one or more of these wonderful fountains.

The best and yet the cheapest.

It certainly affords me great pleasure in being able to offer the best drinking fountain made and at a reasonable price, being cheaper than any other kind that I know of.

It is made of aluminum and screws on an ordinary fruit jar; you can use either quart or half gallon size, as they have both the same size top. We do not furnish the jars, as every one has glass fruit jars. They do not cost over 10 cents each, cheaper than transportation charges. You can use one that you have emptied the fruit from.

Fill the jar with water and then screw on the fountain the same as the lid and turn jar upside down same as in cut; drive three stakes in the ground to keep it from being turned over from any cause. There is not very much danger. I had one standing in one of my yards all summer without any support. I used the half gallon jar.

These fountains are made of aluminum; they are light, strong and will not tarnish, rust or corrode; they will last a lifetime if proper care is given them.

Easy to clean, sanitary and healthy for your birds, easy to fill, easy to handle, and so cheap.

Better throw away your old rusty or wooden drinking fountains and lay in a supply of these automatic aluminum ones; it paid me and I know it will pay you; as I know by actual experience that they are the best thing of the kind ever invented.

We do not furnish the jar, only the aluminum base, which you screw on any ordinary Mason jar.

Price: Each, 35 cents; two, 60 cents; six or more, 30 cents each; charges prepaid.

BERRY'S CHOLERA CURE.

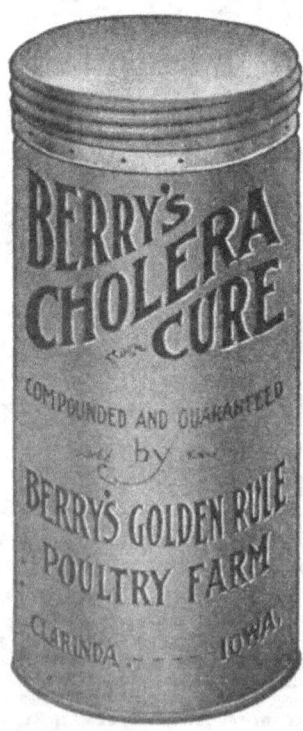

Berry's Automatic Aluminum Drinking Fountains.

This is an excellent remedy and is a safe and certain cure for cholera, dysentery and all other bowel troubles. If your flock has any of the following symptoms, send for a box and it will cure them. Symptoms: Pale combs, which soon become dark; drooping wings, thirst, weakness, frothy discharge resembling in appearance mixed sulphur and water, though sometimes the light yellowish color of the evacuations are absent. Full directions with each package.

Price: Two sizes of cartons, one 50c, and the other $1.00, postpaid.

BERRY'S SURE LICE KILLER AND DISINFECTANT.

This is one of our own preparations and demonstrated by actual experiments in our own chicken houses, that it will do the work. We know we are offering you something that

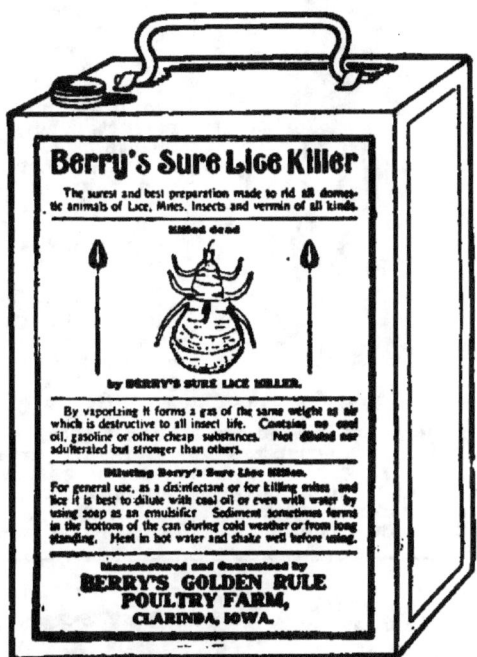

will keep the mites and lice down. It is a liquid and put up in sealed tin cans with screw tops. This insect killer contains more insect destroying qualities than any other preparation ever put on the market for this purpose. The liquid is sprayed or painted on the roosts, nests and sides of the poultry house or on the bedding for hogs. The gas or vapor does the work while the chickens or hogs are asleep. Our price is very reasonable in comparison with others, and very cheap considering the high quality of the preparation. But, in this as all else we offer, we give our customers the advantage of low profits, very shrewd, close buying and low expenses. You cannot afford not to have a can of it to fight the obnoxious mite and lice pests, the worst enemies of chicken raisers.

You pay the freight or express charges. At our prices it will pay you to get up a neighborhood club for it, especially as we make such low prices on five gallons; or if you have a half gallon or gallon can sent with sack of egg, chick food, oyster shell or some bulky produce, by freight, the expense will be nothing.

Price: One-half gallon can, 40c; one gallon can, 75c; two one-gallon cans, $1.40; six one-gallon cans, $3.00. Club in with your neighbor and buy six cans. It will pay.

This is very strong, and much stronger than others, and we know what it will do.

We make this strong guarantee and if Berry's Sure Lice Killer does not kill the mites and lice, we will refund your money. There are no "ifs" and "ands" about it. Money back if it does not do as we say.

BERRY'S ROUP CURE.

Will cure all discharge of the nostrils, rattling of the throat, swelled head, cold or any other symptoms of the roup to which fowls are subject. It is put in the drinking water and the fowls take their own medicine. Roup is one of the worst diseases that the poultrymen have to contend with during the winter months, as quite often they roost in a draught or are exposed so they catch

cold. Roup is only a malignant form of cold, and if taken in time there is no question at all but what it can be cured with our roup remedy. One 50 cent package makes 25 to 30 gallons of the medicine.

Price: Two sized bottles, one 50c and the other $1.00, postpaid.

MARK YOUR PURE BREDS.

For years I have been looking for the best poultry marker made, and I actually believe that I have tried them all. Eureka! I have

found it at last in the PETTEY PUNCH; it is the handiest, simplest, easiest to operate and will do the work better than anything that I have used. It is all steel and nickel plated.

It punches a clean hole of the right size and will not bruise the foot. You cannot miss the web of the foot and punch out the edge, as is common with all other markers.

Everyone that raises pure-bred fowls must have one to perforate the web of the foot so that the birds may be distinguished when they grow up. Price, 25 cents each.

BERRY'S SURE INSECT POWDER.

This is a powder that cannot be excelled for young chicks and matured fowls, also good for any kind of live stock or plants. It kills lice and mites; not to drive them away or put them to sleep for a few hours, as many kinds of powders do, but kills them so they will stay dead. For setting hens and young chicks it is absolutely necessary, as our strong Berry's Sure Lice Killer is too strong for them. But this is just the thing and we use it on Berry's Golden Rule Poultry Farm. It is put up by us and we fully guarantee that if it does not do the work, "your money back." We make good this, as in everything else. You can use one-third of a can and if it does not do what we say it will do, return it and your money will be sent to you.

This preparation is put up in a strong box, in powdered form, and may be applied or dusted on the fowls, eggs or chicks, put in the nest or used on plants wherever vermin have collected.

Price: 25c per 1-lb. box. Add 10c extra if sent by mail.

THE LIGHTNING WHITEWASH SPRAYER AND FORCE PUMP.

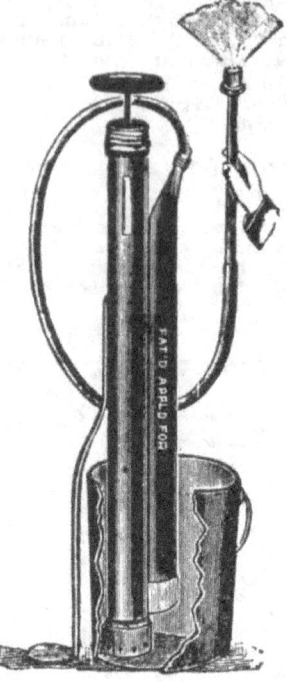

"Cleanliness is next to Godliness," and nowhere is this more true than in the chicken business; cleanliness is absolutely necessary to success in raising and keeping profitable poultry.

We have tried a great many kinds, but always had a little difficulty until we struck the Lightning. It has more advantages and good points in its favor than any other on the market. There is no use of me going into details and pointing out the advantages over all others, but will just simply say that I have tried them to a conclusive satisfaction that this is the sprayer and force pump that you want. Don't buy one of those little tin toys.

This pump is used for a bucket spray and force pump, such as for spraying trees, shrubbery, garden use, vegetable washing, washing wagons, buggies, windows, etc., as well as for all kinds of whitewashing. It will throw a continuous stream 35 feet high.

Each machine is fully guaranteed.

Price complete:
Galvanized iron—
...... $2.50 each
All Brass, $3.50 each

CRUSHED OYSTER SHELLS

Price:

100 lbs. 75c
500 lbs., 60c per 100 lbs.

F. O. B. Clarinda.

MEAT SCRAPS.

Beef scraps are the best of all animal food products. This is a prepared product, cooked and steamed, and let me tell you, if you want something to keep your hens in the best of condition, something to keep up a good circulation, something that makes the combs red, get some beef scraps and mix with a mash; it is also the best known egg food prepared in this way. If you want something to show results and show them quickly, this is the product that will do it.

Price, 50 lb. sack, $1.75; 100 lb. sack, $3.25.

PRACTICAL POULTRY BOOKS

To those who wish to study the Poultry Business further than the information given in this complete book and who desire to specialize on one or two subjects, I have made arrangements to furnish the following list of books at wholesale prices:

Artificial Incubating and Brooding, sent postpaid..$0.50
Eggs and Egg Farming, sent postpaid .... .50
The Chick Book, sent postpaid... .50
The Plymouth Rock (all varieties), sent postpaid...1.00
The Leghorns (all varieties), sent postpaid... 1.00
Turkeys: Their Care and Management, sent postpaid....................................... .75
The Wyandottes (all varieties), sent postpaid... 1.00
The Bantam Fowl, sent postpaid.. .50
The Asiatics—Brahmas, Cochins and Langshans, sent postpaid.............................. .50
Ducks and Geese, sent postpaid.. .75
Poultry House and Fixtures, sent postpaid... .50
Pigeon Queries. A book for pigeon breeders and fanciers. It treats of pigeons from the shell to the show room. Sent postpaid.. .25
999 Questions and Answers on Poultry Raising in all its phases. 132 pages of concisely worded, helpful information. Sent postpaid.. 1.00

Price List for Repairs and Fixtures
FOR BERRY'S BIDDY INCUBATORS AND BROODERS.

In ordering repairs for old machines give us the size of your machines and what year they were shipped from our factory. All items quoted on this list are cash with order.

```
                                                                        Postpaid
Bracket to hold wafer...................................................$0.50
    Side bracket or fulcrum to hold damper lever..........................  .25
    Lamp shelf bracket ..................................................  .50
Burners—No. 20 ..........................................................  .60
    No. 30 ..............................................................  .75
    No. 35 ..............................................................  .75
Burlap for any size chick tray...........................................  .15
Bowls—For No. 20 and No. 35 lamps, each..................................  .60
    For No. 30 lamp, each................................................  .75
Connecting rod and burr .................................................  .25
    (Extra burrs for old rods cannot be furnished.)
Counterweight, used on damper lever......................................  .10
Curtain for any size brooder (price includes chick door curtain).........  .15
Catches for incubator door, each.........................................  .20
Dampers, any size or style...............................................  .15
Egg Testers—Cardboard ...................................................  .25
    Metal, No. 20 .......................................................  .40
    Metal, No. 30 and 35, each...........................................  .45
Funnels, each  ..........................................................  .15
Filler caps, with hoops (state whether for lamp bowl or heater)..........  .20
    Without hoops (state whether for lamp bowl or heater)................  .15
Flues for lamp, any size.................................................  .25
Heater jacket or return draft flue.......................................  .50
Isinglass or mica for lamp flues, any size, two pieces...................  .10
Lamps complete, Special Safety—for incubator and brooder—
    No. 20 and No. 35, each.............................................. 1.25
    No. 30 .............................................................. 1.50
Levers for damper .......................................................  .75
Legs for any size incubator, each........................................  .50
    Set of four legs .................................................... 1.50
Regulator complete ...................................................... 2.00
Screws for burners (to hold flues in place) .............................  .10
Trays—Egg trays, any size, each..........................................  .50
    Chick trays, any size, each..........................................  .40
Thermometers—for incubator ..............................................  .75
Thermometers—for brooder ................................................  .40
Wicks—No. 2 (⅞ in. wide), ½ doz.........................................  .20
    No. 3 (1¼ in. wide), ½ doz. .........................................  .30
    Single wicks, either size, each .....................................  .05
Wafer thermostat, double disc ...........................................  .75
```

To Our Canadian Customers

You no doubt have the impression that if you order thoroughbred poultry and eggs that you will have a high duty to pay. But the Canadian laws admit the shipment of pure-bred poultry and eggs from the United States into your country, free of duty. That is, if such poultry and eggs are shipped for breeding and hatching purposes and if accompanied by manifest properly made out, together with certificate and invoice. We furnish all these necessary papers, so that upon presenting them when claiming the goods, the duty will be released.

A WORD ABOUT DUTY ON INCUBATORS AND BROODERS.

We find that the duty on manufactured articles such as incubators and brooders is reasonably low. We help you out by placing the value of the machine that we sell you on the manifest and certificate down to the lowest figures. The duty is computed by the percent of the given valuation, so we can fix it for you that the duty on these goods is not burdensome.

A WORD ABOUT PARCEL POST AND THE RATES.

We have had one year experimenting with the Parcel Post in egg shipments. We will candidly say that we do not believe that the Postal Department will ever make a success handling goods of a fragile and perishable nature, until such goods are handled free from the sacking system and handled as the Express Companies now handle the goods.

We will admit that in a great many cases the Parcel Post is cheaper, yet I think that it is well worth the difference to know that your eggs will reach you in safety and have not been shaken up so as to put them out of hatching condition. We explain in another part of this book how carefully and securely we pack the eggs for shipment, and surely they must be handled very roughly not to be delivered to you in as good condition as when started from here.

We will ship by Parcels Post if you request; below you will find the rates and also the weight of different baskets of eggs. We urge that you send 10c more than the postage will figure out so that we can have the package insured and if it does not reach you in proper condition a claim can be placed against the department and we can secure an adjustment for you. Clarinda is in the southeastern part of the state of Iowa.

Shipments weigh as follows:—

```
1 setting ................................. 3 lbs.
2 settings ................................ 6 lbs.
3 settings ................................ 8 lbs.
50 to 60 eggs ............................ 11 lbs.
```

PARCEL POST RATES.

Weight	First zone Local rate	Zone rate, 50 miles	Second zone, 50 to 150 miles	Third zone, 150 to 300 miles	Fourth zone, 300 to 600 miles	Fifth zone, 600 to 1000 miles	Sixth zone, 1000 to 1400 miles	Seventh zone, 1400 to 1800 miles	Eighth zone, all over 1800 miles
1 pound	$0.05	$0.05	$0.06	$0.07	$0.08	$0.09	$0.10	$0.11	$0.12
2 pounds	.06	.08	.10	.12	.14	.16	.19	.21	.24
3 pounds	.07	.11	.14	.17	.20	.23	.28	.31	.36
4 pounds	.08	.14	.18	.22	.26	.30	.37	.41	.48
5 pounds	.09	.17	.22	.27	.32	.37	.46	.51	.60
6 pounds	.10	.20	.26	.32	.38	.44	.55	.61	.72
7 pounds	.11	.23	.30	.37	.44	.51	.64	.71	.84
8 pounds	.12	.26	.34	.42	.50	.58	.73	.81	.96
9 pounds	.13	.29	.38	.47	.56	.65	.82	.91	1.08
10 pounds	.14	.32	.42	.52	.62	.72	.91	1.01	1.20
11 pounds	.15	.35	.46	.57	.68	.79	1.00	1.11	1.32

Express Rates

We have found that a good many new customers wish to know just what they will have to pay for express charges. We herewith give you the express rate to the principal points in each state. You can easily approximate what the charges will be. You will readily see that the charges are not high. One setting of eggs will weigh 4 lbs.; 2 settings, 7 lbs. One bird will average 10 lbs. Some of the heavier birds will go over that, but we ship in as light coops as it is possible and make them strong. We do all in our power to make the express and freight charges as low as possible. Poultry supplies by freight.

RATES FROM CLARINDA, IOWA, TO:

	4-lb.	7-lb.	10-lb.	25-lb.
Alabama.				
Birmingham	$0.60	$1.00	$1.10	$1.60
Arkansas.				
Fort Smith	.60	.85	1.30	1.75
Maryland.				
Baltimore	.60	.90	1.00	1.50
Nebraska.				
Alliance	.60	.80	.90	1.30
Hastings	.55	.70	.75	1.10
Lincoln	.40	.50	.55	.75
Omaha	.30	.35	.40	.45
Massachusetts.				
Boston	.60	.90	1.00	1.50
Maine.				
Portland	.60	1.00	1.30	1.85
Arizona.				
Phoenix	.60	1.00	2.05	4.00
California.				
Sacramento	.60	1.10	2.45	4.10
Canada.				
Toronto	.60	.90	1.45	1.95
Connecticut.				
Hartford	.60	1.00	1.10	1.60
Colorado.				
Denver	.60	1.00	1.10	1.60
Delaware.				
Dover	.60	1.00	1.25	1.60
Georgia.				
Macon	.60	1.00	1.15	1.85
Florida.				
Tallahassee	.60	1.00	1.15	1.85
Iowa.				
Albia	.35	.40	.45	.55
Burlington	.40	.50	.55	.75
Council Bluffs	.30	.35	.40	.45
Des Moines	.30	.45	.50	.65
Sioux City	.45	.55	.85	1.00
Illinois.				
Chicago	.50	.60	.70	1.00
Peoria	.45	.55	.60	.85
Quincy	.45	.55	.60	.85
Indiana.				
Evansville	.60	.80	.90	1.30
Ft. Wayne	.60	.75	.80	1.20
Indianapolis	.60	.75	.80	1.20
Idaho.				
Boise	.60	1.10	2.35	2.85
Kentucky.				
Louisville	.60	.80	.90	1.50
Kansas.				
Atchison	.35	.45	.50	.65
Beloit	.60	.75	1.05	1.45
Ft. Scott	.55	.70	.75	1.10
Olathe	.50	.60	.70	1.00
Louisiana.				
New Orleans	.60	1.00	1.15	1.85
Michigan.				
Grand Rapids	.60	.80	.90	1.30
Missouri.				
Bethany	$0.40	$0.50	$0.55	$0.75
Kansas City	.40	.50	.55	.75
St. Joseph	.35	.40	.45	.55
Minnesota.				
Minneapolis	.55	.70	1.10	1.75
Montana.				
Billings	.60	1.00	1.15	1.60
Mississippi.				
Jackson	.60	1.00	1.10	1.60
New Hampshire.				
Concord	.60	1.00	1.35	1.90
Nevada.				
Carson City	.65	1.15	2.60	4.60
North Carolina.				
Raleigh	.60	1.00	1.10	1.85
New York.				
Buffalo	.60	.85	1.00	1.40
New York	.60	.90	1.00	1.50
New Mexico.				
Santa Fe	.60	1.00	1.70	2.60
New Jersey.				
Trenton	.60	.90	1.00	1.50
North Dakota.				
Bismark	.60	1.00	1.55	3.30
Fargo	.60	.85	1.45	2.10
Ohio.				
Toledo	.60	.80	.90	1.30
Oklahoma.				
Oklahoma City	.60	.80	1.20	1.75
Oregon.				
Portland	.60	1.10	2.45	4.10
Pennsylvania.				
Philadelphia	.60	.90	1.00	1.50
Rhode Island.				
Providence	.60	.90	1.00	1.50
South Dakota.				
Sioux Falls	.60	.85	1.20	1.65
Texas.				
Dallas	.60	.90	1.35	1.95
Houston	.60	1.00	1.55	2.15
Virginia.				
Richmond	.60	1.00	1.15	1.85
West Virginia.				
Charleston	.60	1.00	1.15	1.85
Wyoming.				
Cheyenne	.60	1.00	1.10	1.60
Wisconsin.				
Milwaukee	.55	.75	.75	1.00
Washington.				
Walla Walla	.60	1.00	2.00	3.55
Tennessee.				
Nashville	.60	.85	1.00	1.40
South Carolina.				
Columbia	.60	1.00	1.20	2.00
Utah.				
Salt Lake	.60	1.00	2.20	3.20

How to Order

PLEASE READ CAREFULLY.

Order Early. Do not wait too long or until ready to use your cockerel, incubator or eggs unless unavoidable. We will give our best attention to your orders at all times.

We Notify When We Ship. We mail you notice of express shipment or freight receipt when we fill the order.

TERMS.

Cash With Order. We cannot do a credit business as it would take too much extra help in book-keeping and obtaining the standing of those who order, so we could not sell at the prices named in this catalog. It would also delay orders. We are reliable. See reference below.

How to Send Money. Send, at our risk, check on your bank where you have money on deposit, Post Office Order, Registered Letter, Draft or Express Money Order. Very small amounts may be sent in stamps.

REFERENCES.

As references to our honesty and ability to fill your orders and to do as we claim, we refer you by permission to the following business men of Clarinda: C. A. Lisle, Editor Clarinda Herald; Chad. Baker, Agent Adams Express Co.; Wm. Orr, President Clarinda Trust and Savings Bank; J. M. Pierce, of Homestead, Des Moines, Iowa; or any of the agricultural papers.

To Whom It May Concern: December 3rd, 1910.

Mr. and Mrs. A. A. Berry, proprietors of the Berry's Golden Rule Poultry Farm, have been in the pure bred poultry and egg business here for a number of years. I consider them perfectly reliable, and feel that it is their intention to make good every representation they make in regard to the shipments they put out. Yours truly,

A. F. GALLOWAY,
Cashier Clarinda Trust & Savings Bank.

This is to testify that I am well acquainted with the proprietors of Berry's Poultry Farm of this place, and consider them perfectly reliable and you can entrust your orders to them knowing that they will do as they agree. E. G. DAY, Cash. Clarinda Nat. Bank.

Your check is good if you have money in the bank to back it. Many find it the handiest way to remit, therefore I will accept your check, as I have every confidence in you. Please show the same confidence in me by sending in an order.

WHAT WE GUARANTEE.

Our reputation is at stake as to honest dealing and square treatment in every transaction. So, in brief, we guarantee prompt shipments of birds, eggs, incubators, brooders and supplies, quality guaranteed. Safe delivery, lowest expense guaranteed, and in all your dealings you will be treated as our name indicates, on the Golden Rule Plan.

On Fowls.

We guarantee all birds shipped to reach your nearest express office safely and to be exactly as represented. If they are not as represented, will return your money or send you another bird. We use light, strong coops and can ship anywhere. Do not be afraid of distance. We ship anywhere on the globe.

On Eggs.

We guarantee all eggs shipped by us to be first class, from standard bred stock and as represented. We take great pains to have eggs of the highest fertility and know they will hatch if given proper care. In case 60 per cent of eggs shipped are not fertile, and you will say you are sure that it is the fault of the eggs, we will refill the order at half price except when the eggs are sold at reduced price or during July prices.

When eggs are ordered for incubator purposes in quantities, in case of a poor hatch, we will certainly do the fair thing in refilling the order.

OUR INCUBATORS AND BROODERS—IRON CLAD GUARANTEE—THREE HATCH TRIAL.

We guarantee the "Biddy" Incubators and Brooders to be fully as represented. You are to use the "Biddy" three hatches to test it, and we GUARANTEE IT TO GIVE GOOD RESULTS if given a fair trial and the directions followed.

If it does not fill the requirements, notify us, and we will remedy the defect, replace with a new machine without any cost to you, or refund your money.

We also warrant the machine for 5 years.

No strings to this guarantee; we know just what the "Biddy" will do, hence can make such a strong Iron Clad Guarantee.

BASKETS AND PACKING.

We use an extra strong light weight basket made especially for us. We pack securely in excelsior and guarantee the safe arrival of the eggs to any part of the world. Should any damage occur, we will make good, but customers must notify us at once and have notation of the damage written on express or freight receipt by your agent, when you get your goods, in order that we may get your claim for said damages.

Printed by A. B. Morse Co., St. Joseph, Mich.

Questions and Answers

Q. What do you guarantee?

A. The strongest guarantee made by any concern. Full particulars found on page 128.

Q. How do you ship?

A. Full particulars as to ordering and shipping found on pages 127 and 128. **Read it carefully.**

Q. Should eggs be set immediately upon arrival?

A. They should be allowed to rest at least 24 hours before being set under hen or in incubator.

Q. Could you get cheap freight or express rate if I should send the money?

A. No. There is only one rate, which is the cheapest we can secure for our customers. Our name on the package is sufficient inducement for the agent to make the lowest rate possible.

Q. Do you assort your settings? That is, do you send two or more varieties in a setting if desired?

A. You may order as many varieties as described, but in one or two settings we would not advise mixing them up too much for best results.

Q. Is there any way to determine sex of eggs.

A. There is no way of telling what sex an egg will produce. Some claim that round eggs hatch females and the long eggs, males. This, however, is an erroneous idea and cannot be depended upon.

Q. What goes with an incubator?

A. Everything that is necessary for the operation except eggs and oil.

Q. Do you have agents?

A. No. Our margin of profit is very small and we cannot afford to have agents. We sometimes make special prices on large amounts, as for incubators, brooders and supplies in order to get them introduced in a neighborhood. Our catalogue is our only salesman. Our prices are very low, compared with the high quality of birds, eggs, incubators and supplies we sell.

Q. Do you make a special machine for hatching duck and turkey eggs?

A. No. Our Berry's Biddy incubator will hatch them very successfully, using the same trays, but should be hatched by themselves.

Q. What size incubator do you advise using?

A. The 225 egg size is considered the most profitable to buy. It takes but little more oil and no more attention than the smaller one. However, the 115 egg size is a very neat machine and is **the** thing for any one who is not trying to hatch chickens on a large scale. If an accident should occur you do not lose so many eggs.

Q. Would you advise hatching chicks after July 1st?

A. Yes. During July and August some of our customers raise their best hatches. Write us for special summer prices.

Q. Will eggs hatch well that are shipped far?

A. We constantly ship eggs into every state and territory in the United States, also to Canada, Europe, South America and Mexico. In nearly all cases the hatch is satisfactory.

Q. Do you consider water fowls profitable to raise?

A. We undoubtedly do, and there are thousands of farmers who are making lots of money, with very little trouble, raising geese and ducks.

Q. How many fowls can one person take care of?

A. One person can take care of a thousand hens, if he puts in his time as he would at any other work. By the time a person has a thousand hens they are usually able to take things easy and hire a hand to do the work while they manage the business.

Q. When is the best time to set eggs?

A. Early in the spring or any time that suits you. Our rush trade is from January to July. The best months here are April and May, although many raise fall chickens with great success.

Q. What is meant by the double and single mating system?

A. Double mating is a system of using two matings to produce birds of one variety, one mating being to produce the exhibition males and the other the exhibition females. This is practiced by breeders of some parti-colored varieties. Single mating is the mating of standard colored birds of both sexes with the expectation of producing both males and females of standard or exhibition color.

www.ingramcontent.com/pod-product-compliance
Lightning Source LLC
Chambersburg PA
CBHW062324220526
45469CB00008B/2616